I0484389

FEDERAL EXECUTIVE TEAM

Director, Climate Change Science Program ....William J. Brennan

Director, Climate Change Science Program Office ....Peter A. Schultz

Lead Agency Principal Representative to CCSP;
Deputy Under Secretary of Commerce for Oceans and Atmosphere,
National Oceanic and Atmospheric Administration ....Mary M. Glackin

Product Lead, Carnegie Mellon University ....M. Granger Morgan

Synthesis and Assessment Product Advisory
Group Chair; Associate Director, EPA National
Center for Environmental Assessment...Michael W. Slimak

Synthesis and Assessment Product Coordinator,
Climate Change Science Program Office ....Fabien J.G. Laurier

Special Advisor, National Oceanic
and Atmospheric Administration ....Chad A. McNutt

EDITORIAL AND PRODUCTION TEAM

Chair..M. Granger Morgan, Carnegie Mellon Univ.
Scientific Editor ....Jessica Blunden, STG, Inc.
Scientific Editor ....Anne M. Waple, STG, Inc.
Technical Advisor ....David J. Dokken, USGCRP
Production Team Lead .... Sara W. Veasey, NOAA
Graphic Design Co-Lead .... Deborah B. Riddle, NOAA
Designer .... Brandon Farrar, STG, Inc.
Designer .... Glenn M. Hyatt, NOAA
Designer .... Deborah Misch, STG, Inc.
Designer ....Christian Zamarra, STG, Inc.
Copy Editor.. Anne Markel, STG, Inc.
Copy Editor.. Lesley Morgan, STG, Inc.
Copy Editor.. Susan Osborne, STG, Inc.
Copy Editor.. Susanne Skok, STG, Inc.
Copy Editor.. Mara Sprain, STG, Inc.
Copy Editor.. Brooke Stewart, STG, Inc.
Technical Support... Jesse Enloe, STG, Inc.

This Synthesis and Assessment Product, described in the U.S. Climate Change Science Program (CCSP) Strategic Plan, was prepared in accordance with Section 515 of the Treasury and General Government Appropriations Act for Fiscal Year 2001 (Public Law 106-554) and the information quality act guidelines issued by the Department of Commerce and NOAA pursuant to Section 515 <http://www.noaanews.noaa.gov/stories/iq.htm>. The CCSP Interagency Committee relies on Department of Commerce and NOAA certifications regarding compliance with Section 515 and Department guidelines as the basis for determining that this product conforms with Section 515. For purposes of compliance with Section 515, this CCSP Synthesis and Assessment Product is an "interpreted product" as that term is used in NOAA guidelines and is classified as "highly influential". This document does not express any regulatory policies of the United States or any of its agencies, or provide recommendations for regulatory action.

Best Practice Approaches for Characterizing, Communicating, and Incorporating Scientific Uncertainty in Climate Decision Making

Synthesis and Assessment Product 5.2
Report by the U.S. Climate Change Science Program
and the Subcommittee on Global Change Research

Lead Author:
M. Granger Morgan

Contributing Authors:
Hadi Dowlatabadi, Max Henrion, David Keith,
Robert Lempert, Sandra McBride,
Mitchell Small, and Thomas Wilbanks

January 2009,

Members of Congress:

On behalf of the National Science and Technology Council, the U.S. Climate Change Science Program (CCSP) is pleased to transmit to the President and the Congress this Synthesis and Assessment Product (SAP) *Best Practice Approaches for Characterizing, Communicating, and Incorporating Scientific Uncertainty in Decisionmaking*. This is part of a series of 21 SAPs produced by the CCSP aimed at providing current assessments of climate change science to inform public debate, policy, and operational decisions. These reports are also intended to help the CCSP develop future program research priorities.

The CCSP's guiding vision is to provide the Nation and the global community with the science-based knowledge needed to manage the risks and capture the opportunities associated with climate and related environmental changes. The SAPs are important steps toward achieving that vision and help to translate the CCSP's extensive observational and research database into informational tools that directly address key questions being asked of the research community.

The purpose of this SAP is to synthesize and communicate the current state of understanding about the characteristics and implications of uncertainty related to climate change and variability to an audience of policymakers, decision makers, and members of the media and general public with an interest in developing a fundamental understanding of the issue. It was developed with broad scientific input and in accordance with the Guidelines for Producing CCSP SAPs, the Information Quality Act (Section 515 of the Treasury and General Government Appropriations Act for Fiscal Year 2001 (Public Law 106-554)), and the guidelines issued by the Department of Commerce and the National Oceanic and Atmospheric Administration pursuant to Section 515.

We commend the report's authors for both the thorough nature of their work and their adherence to an inclusive review process.

Sincerely,

Carlos M. Gutierrez	Samuel W. Bodman	John H. Marburger III
Secretary of Commerce	Secretary of Energy	Director, Office of Science and
Chair, Committee on Climate Change	Vice Chair, Committee on Climate	Technology Policy
Science and Technology Integration	Change Science and Technology	Executive Director, Committee
	Integration	on Climate Change Science and
		Technology Integration

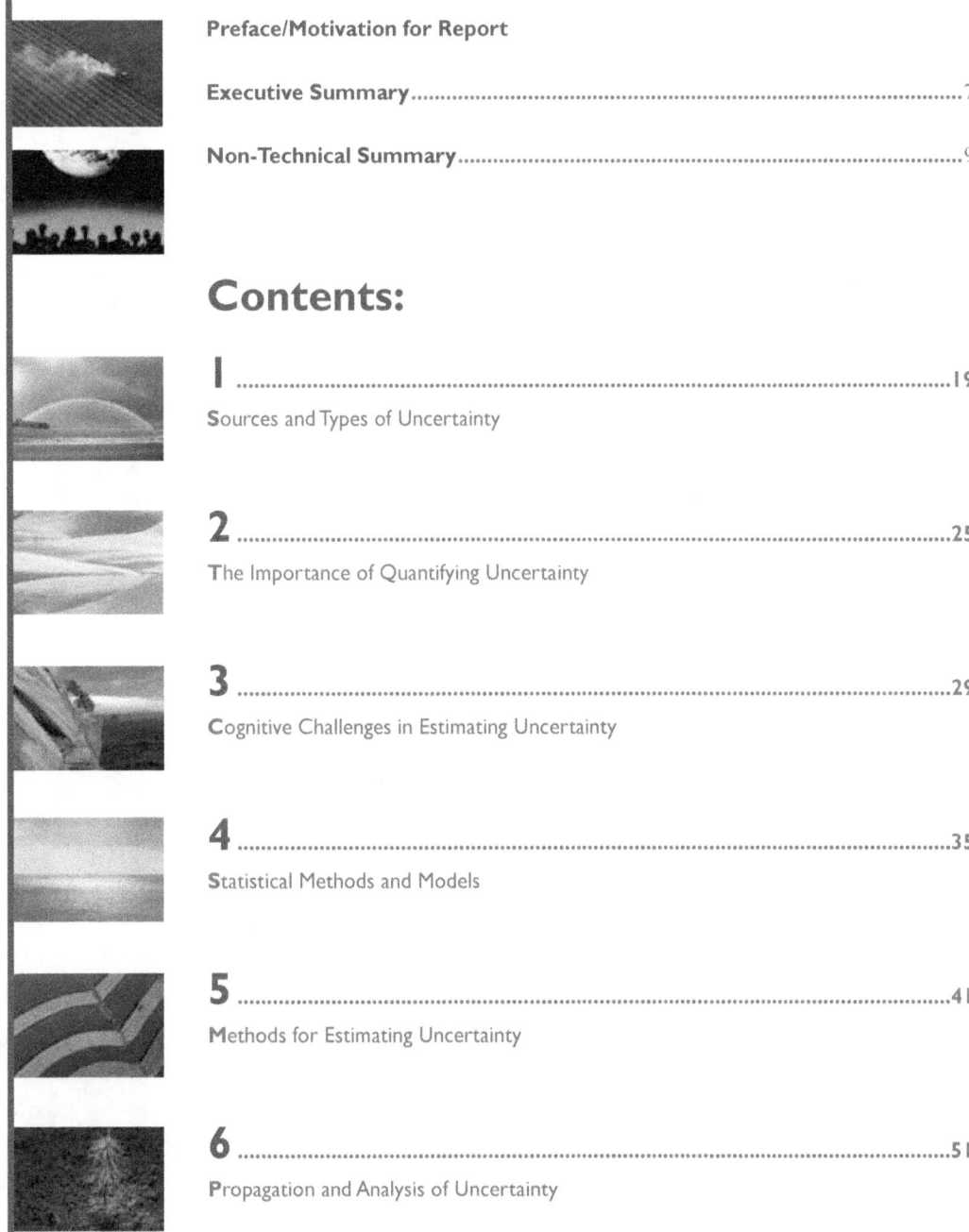

TABLE OF CONTENTS

Contents:

AUTHOR TEAM FOR THIS REPORT

For the entire Report

Lead Author: M. Granger Morgan, Department of Engineering and Public Policy, Carnegie Mellon Univ.

Contributing Authors: Hadi Dowlatabadi, Institute for Resources, Environment and Sustainability, Univ. of British Columbia; Max Henrion, Lumina Decision Systems; David Keith, Department of Chemical and Petroleum Engineering and Department of Economics, Univ. of Calgary; Robert Lempert, The RAND Corporation; Sandra McBride, Duke Univ.; Mitchell Small, Department of Engineering and Public Policy, Carnegie Mellon University; Thomas Wilbanks, Environmental Science Division, Oak Ridge National Laboratory

ACKNOWLEDGEMENTS

Funding for this work was provided by NSF cooperative agreement SES-034578 with the Climate Decision Making Center at Carnegie Mellon Universtiy.

RECOMMENDED CITATION

CCSP, 2009: *Best Practice Approaches for Characterizing, Communicating, and Incorporating Scientific Uncertainty in Decisionmaking.* [M. Granger Morgan (Lead Author), Hadi Dowlatabadi, Max Henrion, David Keith, Robert Lempert, Sandra McBride, Mitchell Small, and Thomas Wilbanks (Contributing Authors)]. A Report by the Climate Change Science Program and the Subcommittee on Global Change Research. National Oceanic and Atmospheric Administration, Washington, DC, 96 pp.

PREFACE

Report Motivation and Guidance for Using this Synthesis/Assessment Report

Lead Author: M. Granger Morgan, Department of Engineering and Public Policy, Carnegie Mellon Univ.

Contributing Authors: Hadi Dowlatabadi, Institute for Resources, Environment and Sustainability, Univ. of British Columbia; Max Henrion, Lumina Decision Systems; David Keith, Department of Chemical and Petroleum Engineering and Department of Economics, Univ. of Calgary; Robert Lempert, The RAND Corporation; Sandra McBride, Duke Univ.; Mitchell Small, Department of Engineering and Public Policy, Carnegie Mellon Univ.; Thomas Wilbanks, Environmental Science Division, Oak Ridge National Laboratory

This Product is one of 21 synthesis and assessment products (SAPs) commissioned by the U.S. Climate Change Science Program (CCSP) as part of an inter-agency effort to integrate federal research on climate change and to facilitate a national understanding of the critical elements of climate change. Most of these products are focused on specific substantive issues in climate science, impacts, and related topics. In contrast, the focus of this Product is methodological.

Uncertainty is ubiquitous. Of course, the presence of uncertainty does not mean that people cannot act. As this Product notes, in our private lives, we decide where to go to college, what job to take, whom to marry, what home to buy, when and whether to have children, and countless other important choices, all in the face of large, and often, irreducible uncertainty. The same is true of decisions made by companies and by governments.

Recent years have seen considerable progress in the development of improved methods to describe and deal with uncertainty. Progress in applying these methods has been uneven, although the field of climate science and impact assessment has done somewhat better than many others.

The primary objective of this Product is to provide a tutorial for the climate analysis and decision-making communities on current best practice in describing and analyzing uncertainty in climate-related problems. While the language is largely semi-technical, much of it should also be accessible to non-expert readers who are comfortable with the treatment of technical topics at the level of journals such as Scientific American.

Because the issue of how uncertainty is characterized and dealt with is of broad importance for public policy, we have also prepared a "Non-Technical Summary". Readers who lack the time or background to read the detailed Product may prefer to start there, and then sample from the main Product as they find topics they would like to learn about in greater depth.

Best Practice Approaches for Characterizing, Communicating, and
Incorporating Scientific Uncertainty in Climate Decision Making

Lead Author: M. Granger Morgan, Carnegie Mellon Univ.

Contributing Authors: Hadi Dowlatabadi, Univ. of British
Columbia; Max Henrion, Lumina Decision Systems; David Keith,
Univ. of Calgary; Robert Lempert, The RAND Corporation; Sandra
McBride, Duke Univ.; Mitchell Small, Carnegie Mellon Univ.; Thomas
Wilbanks, Oak Ridge National Laboratory

This Product begins with a discussion of a
number of formulations of uncertainty and the
various ways in which uncertainty can arise.
It introduces several alternative perspectives
on uncertainty including both the classical or
frequentist view of probability, which defines
probability as the property of a large number
of repeated trials of some process such as the
toss of a coin, and the subjectivist view, in which probability is an indication of degree of belief
informed by all available evidence. A distinction is drawn between uncertainty about the value
of specific quantities and uncertainty about the underlying functional relationships among key
variables. The question of when it is and is not appropriate to represent uncertainty with a
probability distribution is explored. Part 1 of the Product closes with a discussion of "ignorance"
and the fact that while research often reduces uncertainty, it need not always do so; indeed, in
some cases, it may actually lead to greater uncertainty as new unanticipated complexities are
discovered.

Part 2 argues that it is insufficient to describe uncertainty in terms of qualitative language, using
words such as "likely" or "unlikely". Empirical evidence is presented that demonstrates that
such words can mean very different things to different people, or indeed, different things to the
same person in different contexts. Several simple strategies that have been employed to map
words into probabilities in the climate literature are described.

In order to make judgments about, and in the presence of uncertainty, the human mind subconsciously employs a variety of simplified strategies or "cognitive heuristics". In many circumstances, these serve well. However, in some settings, they can lead to significant biases in the
judgments that people make. Part 3 summarizes key findings from the experimental literature
in behavioral decision making, and discusses a number of the cognitive biases that can arise,
including overconfidence, when reasoning and making decisions in the face of uncertainty.

Once uncertainty has been described in a quantitative form, a variety of analytical tools and
models are available to perform analysis and support decision making. Part 4 provides a brief
discussion of a number of statistical models used in atmospheric and climate science. This Part
also discusses methods for hypothesis and model testing as well as a variety of emerging methods
and applications. While the treatment is general, the focus throughout is on climate-related
applications. Box 4.1 provides an illustration of frequentist and Bayesian approaches applied to
the prediction of rainfall.

Part 5 explores two broad methods for estimating uncertainty: model-based approaches and the use of expert judgment obtained through careful systematic "expert elicitation". In both cases illustrations are provided from the climate literature. Issues such as whether and when it is appropriate to combine uncertainty judgments from different experts, and strategies that have been used to help groups of experts develop probabilistic judgments about quantities and model forms, are discussed.

Part 6 explores the issues of how best to propagate uncertainty through models or other decision-making aids, and, more generally, the issues of performing analysis of and with uncertainty. Again, illustrative examples are drawn from the climate literature. Part 7 then explores a range of issues that arise in making decisions in the face of uncertainty, focusing both on classical decision analysis that seeks "optimal strategies", as well as on "resilient strategies" that work reasonably well across a range of possible outcomes, and "adaptive" strategies that can be modified to achieve better performance as the future unfolds. This Part closes with a discussion of deep uncertainty, surprise, and some additional issues related to the discussion of behavioral decision theory, building on ideas introduced in Part 3.

Part 8 addresses a number of issues that arise in communicating about uncertainty, again drawing on the empirical literature in psychology and decision science. Mental model methods for developing communications are outlined. One key finding is that empirical study is absolutely essential to the development of effective communication. With this in mind, there is no such thing as an expert in communication—in the sense of someone who can tell you ahead of time (*i.e.*, without empirical study) how a message should be framed, or what it should say. Part 8 closes with an exploration of the views of a number of leading scientists and journalists who have worked on the difficult problems that arise in communicating about scientific uncertainty.

Finally, Part 9 offers some summary advice. It argues that doing a good job of characterizing and dealing with uncertainty can never be reduced to a simple cookbook. One must always think critically and continually ask questions such as:

- Does what we are doing make sense?
- Are there other important factors that are equally or more important than the factors we are considering?
- Are there key correlation structures in the problem that are being ignored?
- Are there normative assumptions and judgments about which we are not being explicit?
- Is information about the uncertainties related to research results and potential policies being communicated clearly and consistently?

Then, based both on the finding in the empirical literature, as well as the diverse experience and collective judgment of the writing team, it goes on to provide some more specific advice on reporting uncertainty and on characterizing and analyzing uncertainty. This advice can be found on pages 71 through 74.

Lead Author: M. Granger Morgan, Carnegie Mellon Univ.

Contributing Authors: Hadi Dowlatabadi, Univ. of British
Columbia; Max Henrion, Lumina Decision Systems; David Keith,
Univ. of Calgary; Robert Lempert, The RAND Corporation; Sandra
McBride, Duke Univ.; Mitchell Small, Carnegie Mellon Univ.; Thomas
Wilbanks, Oak Ridge National Laboratory

Vaclav Smil (2007), one of the most wide-ranging intellects of our day, observes that "the necessity to live with profound uncertainties is a quintessential condition of our species". Two centuries ago, Benjamin Franklin (1789), an equally wide-ranging intellect of his day, made the identical observation in more colorful and colloquial language when he wrote that "...in this world nothing is certain but death and taxes" and of course, even in that case, the date of one's death and the amount of next year's taxes are both uncertain.

These views about uncertainty certainly apply to many aspects of climate change and its possible impacts, including:

- how the many complex interactions within and among the atmosphere, the oceans, ice in the Arctic and Antarctic, and the living "biosphere" shape local, regional, and global climate;
- how, and in what ways, climate has changed over recent centuries and is likely to change over coming decades;
- how human activities and choices may result in emissions of gases and particles and in changes in land use and vegetation, which together can influence future climate;
- how those changes will affect the climate;
- what impacts a changed climate will have on the natural and human world; and
- how the resulting changes in the natural and human world will feed back on and influence climate in the future.

Clearly the climate system, and its interaction with the human and natural world, is a prime example of what scientists call a "complex dynamic interactive system".

This Product is not about the details of what we know, do not know, could know with more research, or may not be able to know until years after climate has changed, but about these complex processes. These issues are discussed in detail in a number of other reports of the U.S. Climate Science Research Program (CCSP), as well as reports of the Intergovernmental Panel on Climate Change (IPCC), the United States National Research Council, and special studies such as the United States National Assessment, and the Arctic Climate Impact Assessment[1].

---

[1] For access to the various reports mentioned in this sentence, see respectively:
<http://www.climatescience.gov/>;
<http://www.ipcc.ch>; <http://www.nationalacademies.org/publications/>;
<http://www.usgcrp.gov/usgcrp/nacc/default.htm>; and <http://www.acia.uaf.edu/>.

However, for non-technical readers who may not be familiar with the basics of the problem of climate change, we offer a very simple introduction in Box NT.1.

This Product provides a summary of tools and strategies that are available to characterize, analyze, and otherwise deal with uncertainty in characterizing, and doing analysis of, climate change and its impacts. The Product is written to serve the needs of climate scientists, experts assessing the likely impacts and consequences of climate change, as well as technical staff supporting private and public decision makers. As such, it is rather technical in nature, although in most cases we have avoided mathematical detail and the more esoteric aspects of the methods and tools discussed—leaving those to references cited throughout the text.

BOX NT.1: Summary of Climate Change Basics

Carbon dioxide (CO_2) is released to the atmosphere when coal, oil, or natural gas is burned. Carbon dioxide is not like air pollutants such as sulfur dioxide (SO_2), oxides of nitrogen (NO_x), or fine particles. When emissions of these pollutants are stabilized, their atmospheric concentration is also quickly stabilized since they remain in the atmosphere for only a matter of hours or days. The relationship between emissions and concentrations for these pollutants is illustrated in this simple diagram:

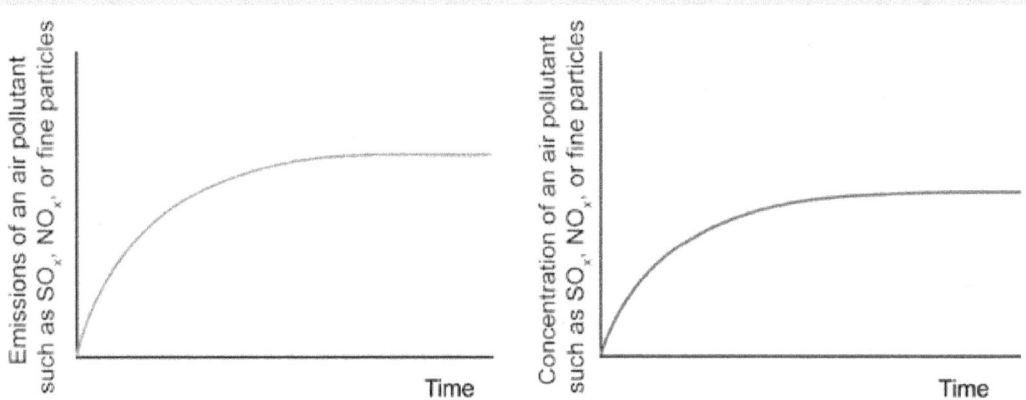

This is not true of carbon dioxide or several other greenhouse gases.

Much of the carbon dioxide that is emitted stays in the atmosphere for over 100 years. Thus, if emissions are stabilized, concentrations will continue to build up, in much the same way that the water level will rise in a bathtub being filled from a faucet that can add water to the tub much faster than a small drain can let it drain out. Again, the situation is summarized in this simple diagram:

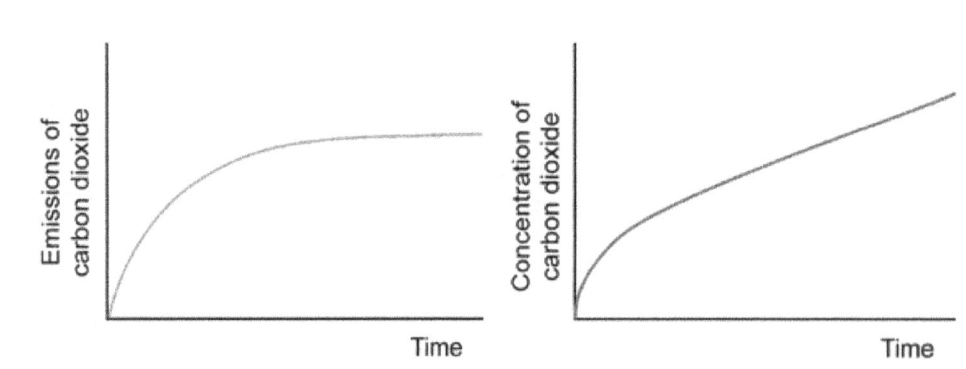

BOX NT.1: Summary of Climate Change Basics *Cont'd*

In order to stabilize atmospheric concentrations of carbon dioxide, worldwide emissions must be dramatically reduced (most experts would say by something like 70 to 90 percent from today's levels depending on the assumptions made about the processes involved and the concentration level that is being sought). Again, here is a simple diagram:

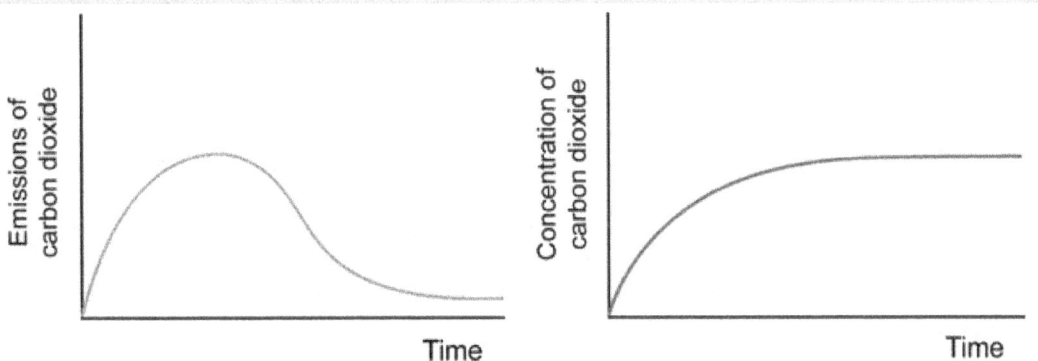

Summarizing, there are three key facts that are important to understand to be an informed participant in policy discussions about climate change:

- When coal, oil, and natural gas (*i.e.*, fossil fuels) are burned or land is cleared or burned, CO_2 is created and released into the atmosphere. There is *no* uncertainty about this.
- Because CO_2 (and other greenhouse gases) trap heat, if more is added to the atmosphere, warming will result that can lead to climate change. Many of the details about how much warming, how fast, and similar issues *are* uncertain.
- CO_2 (and other greenhouse gases) are not like conventional air pollution such as SO_2, NO_x, or fine particles. Much of the CO_2 that enters the atmosphere remains there for more than 100 years. In order to reduce concentration (which is what causes climate change), emissions must be dramatically reduced. There is no uncertainty about this basic fact, although there is uncertainty about how fast and by how much emissions must be reduced to achieve a specific stable concentration. Most experts would suggest that a reduction of CO_2 emissions of between 70 and 90 percent from today's levels is needed. This implies the need for dramatic changes in energy and other industrial systems all around the globe.

This Product explores eight aspects of this topic. Then, in Part 9, the Product concludes with some guidance for researchers and policy analysts that is based both on relevant scientific literature and on the diverse experience and collective judgment of the writing team.

PART 1: SOURCES AND TYPES OF UNCERTAINTY

Uncertainty arises in a number of ways and for a variety of reasons. First, and perhaps simplest, is uncertainty in measuring specific quantities, such as temperature, with an instrument, such as a thermometer. In this case, there can be two sources of uncertainty.

The first is random errors in measurement. For example, if you and a friend both look at a typical backyard thermometer and record the temperature, you may write down slightly different numbers because the two of you may read the location of the red line just a bit differently. Similar issues arise with more advanced scientific instruments.

The second source of uncertainty that may occur involves a "systematic" error in the measurement. Again, in the case of the typical backyard thermometer, perhaps the company that printed the scale next to the glass didn't get it on in just the right place, or perhaps the glass slid a bit with respect to the scale. This could result in all the measurements that you and your friend write down being just a bit high or low, and, unless you checked your thermometer against a very accurate one (*i.e.*, "calibrated" it), you'd never know this problem existed. Again, similar issues can arise with more advanced scientific instruments. Errors can also result in the recording, reporting, and archiving of measurement data.

Beyond random and systematic measurement errors lies a much more complicated kind of potential uncertainty. Suppose, for example, you want to know how much rain your garden will receive next summer. You may have many years of data on how much rain has fallen in your area during the growing season, but, of course, there will be some variation from year to year and from place to place. You can compute the average of past measurements, but if you want to have an estimate for *next* summer at a specific location, the average does not tell you the whole story. In this case, you will want to look at the distribution of the amounts that fell over the years, and figure out the odds that you will get varying amounts by examining how often that amount occurred in the past. If the place where the rain gauge is located gets a different amount of rain than the amount your garden gets, you'll also need to factor that in.

Continuing with this example, if the sum of all rainfall in your region is gradually changing over the years (either because of natural long-term variability or because of systematic climate change), using the distribution of past rainfall will not be a perfect predictor of future rainfall. In this case, you will also need to look at (or try to predict) the trend over time.

Suppose that you want to know the odds that there will be more rain than 45 inches, and suppose that over the past century, there has been only one growing season in which there has been more than that much rain. In this case, since you don't have enough data for reliable statistics, you will have talk to experts (and perhaps have them use a combination of models, trend data, and expert judgment) to get you an estimate of the odds.

Finally, suppose (like most Americans, the authors included) you know nothing about sumo wrestling, but you need to know the odds that a particular sumo wrestler will win the next international championship. In this case, your best option is probably to carefully interview a number of the world's leading sumo coaches and sports commentators and "elicit" odds from each of them. Analysts often do very similar things when they need to obtain odds on the future value of specific climate quantities. This process is known as "expert elicitation". Doing it well takes careful preparation and execution. Results are typically in the form of distributions of odds called "probability distributions".

All of these examples involve uncertainty about the value of some quantity such as temperature or rainfall. There can also be uncertainty about how a physical process works. For example, before Isaac Newton figured out the law of gravity, which says the attraction between two masses (like the Sun and the Earth; or an apple and the Earth) is proportional to the product of the two masses and inversely proportional to the square of the distance between them, people were uncertain about how gravity worked. However, they certainly knew from experience that something like gravity existed. We call this kind of uncertainty "model uncertainty". In the context of the climate system, and the possible impacts of climate change, there are many cases where we do not understand all the physical, chemical, and biological processes that are involved—that is, there are many cases in which we are uncertain about the underlying "causal model". This type of uncertainty is often more difficult to describe and deal with than uncertainty about the value of specific quantities, but progress is being made on developing methods to address it.

Finally, there is ignorance. For example, when Galileo Galilei first began to look at the heavens through his telescope, he may have had an inkling that the Earth revolved around the Sun, but he had no idea that the Sun was part of an enormous galaxy, and that our galaxy was just one of billions in an expanding universe. Similarly, when astronomers built the giant 200-inch telescope on Mount Palomar, they had no idea that at the center of our galaxy lay a massive "black hole". These are examples of scientific ignorance. Only as we accumulate more and more evidence that the world does

In the context of the climate system, and the possible impacts of climate change, there are many cases where we do not understand all the physical, chemical, and biological processes that are involved.

not seem to work exactly as we think it does, do scientists begin to get a sense that perhaps there is something fundamental going on that they have not previously recognized or appreciated. Modern scientists are trained to keep looking for indications of such situations (indeed, that's what wins Nobel prizes) but even when a scientist is looking for such evidence, it may be very hard to see, since all of us, scientists and non-scientists alike, view the world through existing knowledge and "mental models" of how things around us work. There may well still be a few things about the climate system, or climate impacts, about which we are still completely ignorant—and don't even know to ask the right questions.

While Donald Rumsfeld (2002) was widely lampooned in the popular press, he was absolutely correct when he noted that "…there are known unknowns. That is to say, we know there are some things we do not know. But there are also unknown unknowns, the ones we don't know we don't know". But perhaps the ever folksy but profound Mark Twain put it best when he noted, "It ain't what you don't know that gets you in trouble. It's what you know for sure that just ain't so"[2].

PART 2: THE IMPORTANCE OF QUANTIFYING UNCERTAINTY

In our day-to-day discussion, we use words to describe uncertainty. We say:
> "I think it is very likely she will be late for dinner".
> "I think it is unlikely that the Pittsburgh Pirates will win next year's World Series".
> "I'll give you even odds that he will or will not pass his driver's test".
> "They say nuclear war between India and Pakistan is unlikely next year".
> "The doctor says that it is likely that the chemical TZX causes cancer in people".

People often ask, "Why not just use similar words to describe uncertainty about climate change and its impacts?"

Experimental studies have found that such words can mean very different things to different people. They can also mean very different things to the same person in different situations.

Think about betting odds. Suppose that to one person "unlikely" means that they think there is only 1 chance in 10 that something will happen, while to another person the same word means they think there is only one chance in a thousand that that same thing will happen. In some cases, this difference could be very important. For example, in the second case, you might be willing to make a big investment in a company if your financial advisor tells you they are "unlikely" to go bankrupt—that is, the odds are only 1 in 1,000 that it will happen. On the other hand, if by unlikely the advisor actually means a chance of 1 in 10, you might not want to put your money at risk.

The same problem can arise in scientific communication. For example, some years ago members of the U.S. Environmental Protection Agency (EPA) Science Advisory Board were asked to attach odds to the statement that a chemical was "likely" to cause cancer in humans or "not likely" to cause cancer in humans. Fourteen experts answered these questions. The odds for the word "likely" ranged from less than 1 in 10 down to about 1 in 1,000! The range was even wider for the odds given on the word "not likely". There was even an overlap…where a few experts used the word "likely" to describe the same odds that other experts described as "not likely".

Because of results like this, it is important to insist that when scientists and analysts talk about uncertainty in climate science and its impacts, they tell us in quantitative terms what they mean by the uncertainty words they use. Otherwise, nobody can be sure of what they are saying.

The climate community has been better than a number of other communities (such as environmental health) in doing this. However, there is still room for improvement. In the final Part of this Product, the authors offer advice on how they think this should best be done.

There may well still be a few things about the climate system, or climate impacts, about which we are still completely ignorant—and don't even know to ask the right questions.

[2] < http://www.quotedb.com/quotes/1097>.

PART 3: COGNITIVE CHALLENGES IN ESTIMATING UNCERTAINTY

Humans are very good at thinking about and doing lots of things. However, experimental psychologists have found that the way our brains make some judgments, such as those involved in estimating and making decisions about uncertainty, involves unconsciously using some simple rules. These simple rules (psychologists call them "cognitive heuristics") work pretty well most of the time. However, in some circumstances they can lead us astray.

For example, suppose I want to estimate the odds that when I drive to the airport tomorrow morning, I'll see a state police patrol car. I have made that trip at that time of day many times in the past. So, unless there is something unusual going on tomorrow morning, the ease with which I can imagine encountering a state police car on previous trips will probably give me a pretty good estimate of the odds that I'll see one tomorrow.

However, suppose that, instead, I had to drive to the airport tomorrow at 3:30 a.m. I've never done that before (and hope I'll never have to do it). However, if I try to estimate the odds of encountering a state police car on that trip, experience from previous trips, or my imagination about how many state police may be driving around at that time of night, may not give me a very accurate estimate.

This strategy that our minds use subconsciously to estimate probabilities in terms of how easily we can recall past events or circumstances, or imagine them in the future, is a "cognitive heuristic" called "availability". We make judgments in terms of how available experience or imagination is when our minds consider an issue of uncertainty.

Part 3 of the Product describes several such cognitive heuristics. The description is largely non-technical so readers who find these issues interesting should find they could read this part of the Product without much difficulty.

The other issue discussed in Part 3 is overconfidence. There is an overwhelming amount of

evidence from dozens of experimental studies done by psychologists and by decision analysts, that when people judge how well they know an uncertain quantity, they set the range of their uncertainty much too narrowly.

For example, suppose you ask a whole bunch of your adult friends how high Mt. McKinley in Alaska is, or how far it is between Philadelphia and Pittsburgh. But you don't ask them just for their best guess. You ask them for a range. That is, you say, "give me a high estimate and a low estimate of the distance in miles between Philadelphia and Pittsburgh such that there are only 2 chances in 100 that the real distance falls outside of that range". Sounds simple, but when thousands of people have been asked thousands of questions like this, and their uncertainty range is compared with the actual values of the answers, the real answers fall outside of the range they estimated much more than two percent of the time (indeed, sometimes as much as almost half the time).

What does this mean? It means that we all tend to be overconfident about how well we know things that we know are uncertain. And, it is not just ordinary people making judgments about ordinary things such as the weight of bowling balls or the distance from Philadelphia to Pittsburgh. Experts have the same problem.

What does all this have to do with climate change? It tells us that when scientists make estimates of the value of uncertain quantities, or when they, or decision makers, make judgments about uncertain science involving climate change and its impacts, these same processes will be operating. We can't completely get rid of the biases created by cognitive heuristics, nor can we completely eliminate overconfidence. But if we are aware of these tendencies, and the problems they can lead to, we may all be able to do a better job of trying to minimize their impacts.

PART 4: STATISTICAL METHODS AND MODELS

Statistical methods and models play a key role in the interpretation and synthesis of observed climate data and the predictions of numerical climate models. This Part provides a summary

There is an overwhelming amount of evidence from dozens of experimental studies done by psychologists and by decision analysts, that when people judge how well they know an uncertain quantity, they set the range of their uncertainty much too narrowly.

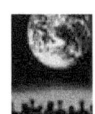

of some of the statistical methods being used for climate assessment, including procedures for detecting longer-term trends in noisy records of past climate that include year-to-year variations as well as various more periodic fluctuations. Such methods are especially important in addressing the question, "What long-term changes in climate are occurring?"

This Part also discusses a number of other issues, such as methods to assess how well alternative mathematical models fit existing evidence. Methods for hypothesis testing and model selection are presented, and emerging issues in the development of statistical methods are discussed.

Rather than give a detailed technical tutorial, this Part focuses on identifying key strategies and analytical tools, and then referring expert readers to relevant review articles and more detailed technical papers.

Many non-technical readers will likely find much of the discussion in this Part too detailed to be of great interest. However, many may find it useful to take a look at Box 4.1 "Predicting Rainfall: An Illustration of Frequentist and Bayesian Approaches" that appears at the end of Part 4. The problems of developing probabilistic descriptions (or odds) on the amount of future rainfall in some location of interest are discussed, first in the presence of various random and periodic changes (wet spells and dry spells) and then in the more complicated situation in which climate change (a long-term trend) is added.

PART 5: METHODS FOR ESTIMATING UNCERTAINTY

Many of the facts and relationships that are important to understanding the climate system and how climate may change over the coming decades and centuries will likely remain uncertain for years to come. Some will probably not be resolved until substantial changes have actually occurred.

While a variety of evidence can be brought to bear to gain insight about these uncertainties, in most cases no single piece of evidence or experimental result can provide definitive

answers. Yet research planners, groups attempting to do impact assessment, policy makers addressing emissions reductions, public and private parties making long-lived capital investment decisions, and many others, all need some informed judgment about the nature and extent of the associated uncertainties.

Two rather different strategies have been used to explore the nature of key uncertainties about climate science, such as the amount of warming that would result if the concentration of carbon dioxide in the atmosphere is doubled and then held constant (this particular quantity is called the "climate sensitivity").

The first section of Part 5 discusses a number of different ways in which climate models have been used in order to gain insight about, and place limits on, the amount of uncertainty about key aspects of the climate system. Some of these methods combine the use of models with the use of expert judgments.

The second section of Part 5 discusses issues related to obtaining and using expert judgments in the form of probability distributions (or betting odds) from experts on what a key value might be, based on their careful consideration and synthesis of all the data, model results, and theoretical arguments in the literature. Several figures in the latter part of this discussion show illustrations of the types of results that can be obtained in such studies. One of the interesting findings is that when these methods are used with individual experts, the resulting impression of the overall level of uncertainty appears to be somewhat greater (that is, the spread of the distributions is somewhat wider) than the results that emerge from consensus panels such as those of the IPCC.

PART 6: PROPAGATION AND ANALYSIS OF UNCERTAINTY

Probabilistic descriptions of what is known about key quantities, such as how much warmer it will get as the atmospheric concentration of carbon dioxide rises or how much the sea level will increase as the average temperature of the Earth increases, can have value in their own right as an input to research planning and in a variety of assessment activities. Often, however,

Many of the facts and relationships that are important to understanding the climate system and how climate may change over the coming decades and centuries will likely remain uncertain for years to come.

There are a number of things about climate change and its likely consequences that are unique. However, uncertainty, even irreducible uncertainty, is not one of them.

analysts want to incorporate such probabilistic descriptions in subsequent modeling and other analyses. Today, this is usually done by running the analysis over and over again on a fast computer, using different input values, from which it is possible to compile the results into probability distributions. This approach is termed "stochastic simulation". Today, a number of standard software tools are available to support such analysis.

Some climate analyses use a single model to estimate what decision or policy is "optimal" in the sense that it has the highest "expected value" (*i.e.*, offers the best bet). However, others argue that because the models used in such analyses are themselves uncertain, it is not wise to search for a single "optimal" answer; it is better to search for answers or policies that are likely to yield acceptable results across a wide range of models and future outcomes. Part 6 presents several examples of results from such analysis.

PART 7: MAKING DECISIONS IN THE FACE OF UNCERTAINTY

There are a number of things about climate change and its likely consequences that are unique. However, uncertainty, even irreducible uncertainty, is not one of them. In our private lives, we decide where to go to college, what job to take, whom to marry, what home to buy, when and whether to have children, and countless other important choices, all in the face of large, and often irreducible, uncertainty. The same is true of decisions made by companies and by governments.

A set of ideas and analytical methods called "decision analysis" has been developed to assist in making decisions in the face of uncertainty. If one can identify the alternatives that are available, identify and estimate the probability of key uncertain events, and specify preferences (utilities) among the range of possible outcomes, these tools can provide help in framing and analyzing complex decisions in a consistent and rational way. Decision analysis has seen wide adoption by private sector decision makers—such as major corporations facing difficult and important decisions. While more controversial, such analysis has also seen more

limited application to public sector decision making, especially in dealing with more technocratic issues.

Of course, even if they want to, most people do not make decisions in precise accordance with the norms of decision analysis. A large literature, based on extensive empirical study, now exists on "behavioral decision theory". This literature describes how and why people make decisions in the way that they do, as well as some of the pitfalls and contradictions that can result. Part 8 provides a few brief pointers into that literature, but does not attempt a comprehensive review. That would require a paper at least as long as this one.

For both theoretical and practical reasons, there are limits to the applicability and usefulness of classic decision analysis to climate-related problems. Two strategies may be especially appealing in the face of high uncertainty:

- Resilient Strategies: In this case, the idea is to try to identify the range of future circumstances that one might face, and then seek to identify approaches that will work reasonably well across that range.
- Adaptive Strategies: In this case, the idea is to choose strategies that can be modified to achieve better performance as one learns more about the issues at hand and how the future is unfolding.

Both of these approaches stand in sharp contrast to the idea of developing optimal strategies that has characterized some of the work in the climate change integrated assessment community, in which it is assumed that a single model reflects the nature of the world with sufficient accuracy to be the basis for decision making and that the optimal strategy for the world will be chosen by a single decision maker.

The "precautionary principle" is another decision strategy often proposed for use in the face of high uncertainty. There are many different notions of what this approach does and does not entail. In some forms, it incorporates ideas of resilient or adaptive policy. In some forms, it can also be shown to be entirely constant with a decision analytic problem framing. Precaution is often in the eye of the beholder. Thus, for example, some have argued that while the European Union has been more precautionary with respect to CO_2 emissions in promoting the wide adoption of fuel efficient diesel automobiles, the United States has been more precautionary with respect to health effects of fine particulate air pollution, stalling the adoption of diesel automobiles until it was possible to substantially reduce their particulate emissions.

PART 8: COMMUNICATING UNCERTAINTY

Many technical professionals have argued that one should not try to communicate about uncertainty to non-technical audiences. They suggest laypeople won't understand and that decision makers want definitive answers—that is, advice from what are often referred to as "one armed scientists"[3].

We do not agree. Non-technical people deal with uncertainty, and statements of probability, all the time. They don't always reason correctly about probability, but they can generally get the gist (Dawes, 1988). While they may make errors about the details, people, for the most part, manage to deal with probabilistic weather forecasts about the likelihood of rain or snow, point spreads at the track, and similar probabilistic information. The real issue is to frame things in familiar and understandable terms.

When should probability be communicated in terms of odds (the chance that the Pittsburgh Pirates will win the World Series this year is about 1 in 100) or in terms of probabilities (the probability that the Pittsburgh Pirates will win the World Series this year is 0.01[4])? Psychologist Baruch Fischhoff and colleagues (2002) suggest that:
- Either will work, if they're used consistently across many presentations.
- If you want people to understand one fact, in isolation, present the result both in terms of odds and probabilities.

- In many cases, there's probably more confusion about what is meant by the specific events being discussed than about the numbers attached to them.

Part 8 briefly discusses some empirical methods that can be used to develop and evaluate understandable and useful communications about uncertain technical issues for non-technical and semi-technical audiences. This approach uses "mental model" methods to learn in some detail what people know and need to know about the topic. Then, having developed a pilot communication working with members of the target audience, the message is extensively tested and refined until it is appropriately understood. One key finding is that empirical study is absolutely essential to the development of effective communication. With this in mind, there is no such thing as an expert in communication—in the sense of someone who can tell you ahead of time (*i.e.*, without empirical study) how a message should be framed, or what it should say.

The presence of high levels of uncertainty offers people who have an agenda with an opportunity to "spin the facts". In addition, many reporters are not in a position to make their own independent assessment of the likely accuracy of scientific statements, seek conflict, and report the views of those holding widely divergent views in just a few words and with very short deadlines. Thus, it is not surprising that the issue of climate change and its associated uncertainties has presented particularly challenging issues for members of the press who are trying to cover the issue in a balanced and responsible way.

In an environment in which there is high probability that many statements a scientist makes about uncertainties will immediately be seized upon by advocates in an ongoing public debate, it is perhaps understandable that many scientists choose to just keep their heads down, do their research, and limit their communication to publication in scientific journals and presentations at professional scientific meetings.

While we do not reproduce it here, the latter portion of Part 8 contains some thoughtful reflection on these issues from several leading scientists and members of the press.

---

[3] The reference, of course, being to experts who always answered his questions "on the one hand...but on the other hand...," the phrase is usually first attributed to Senator Edmund Muskie.

[4] Strictly odds are defined as p/(1-p) but when p is small, the difference between odds of 1 in 99 and 1 in 100 is often ignored when presenting results to non-technical audiences.

> Many technical professionals have argued that one should not try to communicate about uncertainty to non-technical audiences. We do not agree. Non-technical people deal with uncertainty, and statements of probability, all the time. They don't always reason correctly about probability, but they can generally get the gist.

PART 9: SOME SIMPLE GUIDANCE FOR RESEARCHERS

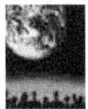

The final Part of the Product provides some advice and guidance to practicing researchers and policy analysts who must address and deal with uncertainty in their work on climate change, impacts, and policy.

However, before turning to specific recommendations, this Part begins by reminding readers that doing a good job of characterizing and dealing with uncertainty can never be reduced to a simple cookbook. Researchers and policy analysts must always think critically and continually ask themselves questions such as:

- Does what we are doing make sense?
- Are there other important factors that are equally or more important than the factors we are considering?
- Are there key correlation structures in the problems that are being ignored?
- Are there normative assumptions and judgments about which we are not being explicit?
- Is information about the uncertainties related to research results and potential policies being communicated clearly and consistently?"

The balance of the final Part provides specific guidance to help researchers and analysts to do a better job of reporting, characterizing, and analyzing uncertainty. Some of this guidance is based on available literature. However, because doing these things well is often as much an art as it is a science, the recommendations also draw on the very considerable and diverse experience and collective judgment of the writing team.

Rather than reproduce these recommendations here, we refer readers to the discussion at the end of Part 9.

PART 1

Sources and Types of Uncertainty[1]

Lead Author: M. Granger Morgan, Carnegie Mellon Univ.

Contributing Authors: Hadi Dowlatabadi, Univ. of British Columbia; Max Henrion, Lumina Decision Systems; David Keith, Univ. of Calgary; Robert Lempert, The RAND Corporation; Sandra McBride, Duke Univ.; Mitchell Small, Carnegie Mellon Univ.; Thomas Wilbanks, Oak Ridge National Laboratory

There are a number of things about climate change and its likely consequences that are unique. However, uncertainty, even irreducible uncertainty, is not one of them. Uncertainty is ubiquitous in virtually all fields of science and human endeavor. As Benjamin Franklin wrote in 1789 in a letter to Jean-Baptiste Leroy, "...in this world nothing is certain but death and taxes". And, even in these cases, the timing and nature of the events are often uncertain.

Sometimes uncertainty can be reduced through research, but there are many settings in which one simply cannot resolve all important uncertainties before decisions must be made. In our private lives, we choose where to go to college, what career to pursue, what job to take, whom to marry, whether and when to have children, all in the face of irreducible uncertainty. Similarly, corporations and governments regularly choose what policies to adopt, and where to invest resources, in the face of large and irreducible uncertainty.

By far, the most widely used formal language of uncertainty is probability[2]. Many of the ideas and much of the vocabulary of probability were first developed in a "frequentist" framework to describe the properties of random processes, such as games of chance, that can be repeated many times. In this case, assuming that the process of interest is stable over time, or "stationary", probability is the value to which the event frequency converges in the long run as the number of trials increases. Thus, in this frequentist or classical framework, probability is a property of a theoretically infinite series of trials, rather than of a single event.

> By far, the most widely used formal language of uncertainty is probability.

While today some people stick to a strict classical interpretation of probability, many statisticians, as well as many of the experimental scientists we know, often adopt a "personalist", "subjectivist", or "Bayesian" view. In many settings, this has the consequence that probability can be used as a statement of a person's degree of belief given all available evidence. In this formulation, probability is not only a function of an event, but also of the state of information i that is available to the person making the assessment. That is, the probability, P, of event X is represented as $P(X|i)$

---

[1] Portions of the discussion in this Part draw heavily on ideas and language from Morgan and Henrion (1990).

[2] There are a few alternative "languages" that have been advanced to describe and deal with uncertainty. These are briefly discussed in Part 2.

where the notation "|i" reads "conditional on i". Thus, $P(X|i)$ means the probability given that all the information is available to the person making the judgment at the same time when the value of the probability P is made. In this framework, obviously a person's value of P may change as more or different information, i, becomes available.

In a personalist or Bayesian framework, it is perfectly appropriate to say, based on a subjective interpretation of polling data, results from focus group discussions, and one's own reading of the political climate, "I think there is an 80 percent chance that Jones will win the next congressional election in this district". However, because it involves the outcome of a single unique future event, such a statement has no meaning in a frequentist framework.

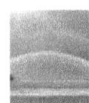

Subjective probabilities—a statement of a person's degree of belief given all available evidence— are intended to characterize the full spectrum of degrees of belief one might hold about uncertain propositions.

In the face of large amounts of data on a repeating event, and a belief that the process being considered is stationary, the subjectivist probability should reduce to the same value as the classical probability. Thus, for example, if you need to estimate the probability that the mid-morning high speed Shinkansen train from Kyoto will arrive on time in Tokyo on a Tuesday morning next month, and you have access to a dataset of all previous arrival times of that train, you would probably want to simply adopt the histogram of those times as your probability distribution on arrival time.

Suppose, however, that you want to estimate how long it takes to complete the weekly shopping for a family of four in your community. If you happen to be the person doing the shopping for a family of four on a regular basis in that community, then, as in the case with the Shinkansen, you will have hundreds of observations to rely on in estimating a probability distribution. The large amount of data available to you helps you understand that the answer has features that depend on the time of day, day of the week, special occasions, and so on. If you do not shop that often, your ability to estimate time for shopping will be less informed and more likely to be in error.

Does a subjectivist view mean that one's probability can be completely arbitrary? "No", Morgan and Henrion (1990) answer, "...because

if they are legitimate probabilities, they must be consistent with the axioms of probability. For example, if you assign probability p that an event X will occur, you should assign 1-p to its complement that X doesn't occur. The probability that one of a set of mutually exclusive events occurs should be the sum of their probabilities. In fact, subjective probabilities should obey the same axioms as objective or frequentist probabilities, otherwise they are not probabilities..."

Subjective probabilities are intended to characterize the full spectrum of degrees of belief one might hold about uncertain propositions. However, there exists a long-standing debate as to whether this representation is sufficient. Some judgments may be characterized by a degree of ambiguity or imprecision distinct from estimates of their probability. Writing about financial matters, Knight (1921) contrasted risk with uncertainty, using the first term to refer to random processes whose statistics were well known and the latter term to describe unknown factors poorly described by quantifiable probabilities. Ellsberg (1961) emphasized the importance of this difference in his famous paradox, where subjects are asked to play a game of chance in which they do not know the probabilities underlying the outcomes of the game[3]. Ellsberg found that many subjects make choices that are inconsistent with any single estimate of probabilities, which nonetheless reflect judgments about which outcomes can be known with the most confidence.

Guidance developed by Moss and Schneider (2000) for the IPCC on dealing with uncertainty describes two key attributes that they argue are important in any judgment about climate change: the amount of evidence available to support the judgment being made and the degree of consensus within the scientific

---

[3] Specifically consider two urns each with 100 balls. In urn 1, the color ratio of red and blue balls is not specified. Urn 2 has 50 red and 50 blue balls. If asked to bet on the color of a ball drawn from one of these urns, most people do not care if the ball is drawn from urn 1 or 2 and give a probability to either color of 0.5. However, when asked to choose an urn when betting on a specified color, most people prefer urn 2. The first outcome implies $p(r_1)=p(r_2)=p(b_1)=p(b_2)$, while the second, it is argued, implies $p(r_1)<p(r_2)$ and $p(b_1)<p(b_2)$. Ellsberg and others discuss this outcome as an illustration of an aversion to ambiguity.

State of knowlege is:

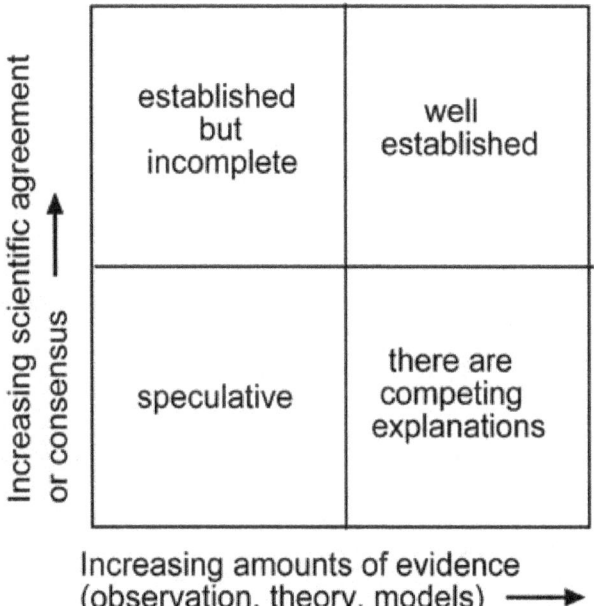

Figure 1.1 Categorization of the various states of knowledge that may apply in different aspects of climate and related problems. Redrawn from Moss and Schneider (2000). *The Guidance Notes for Lead Authors of the IPCC Fourth Assessment Report on Addressing Uncertainties* (IPCC, 2005) adopted a slightly modified version of this same diagram.

community about that judgment. Thus, they argue, judgments can be sorted into four broad types as shown in Figure 1.1[4]. Many decisions involving climate change entail judgments in all four quadrants of this diagram.

Subjective probabilities seem clearly appropriate for addressing the established cases across the top of this matrix. There is more debate about the most appropriate methods for dealing with the others. A variety of approaches exist, such as belief functions, certainty factors, second order probabilities, and fuzzy sets and fuzzy logic, that attempt to quantify the degree of belief in a set of subjective probability judgments[5]. Each of these approaches provides an alternative calculus that relaxes the axioms of probability. In particular, they try to capture the idea that one can gain or lose confidence in one of a mutually exclusive set of events without necessarily gaining or losing confidence in

the other events. For instance, a jury in a court of law might hear evidence that makes them doubt the defendant's alibi without necessarily causing them to have more confidence in the prosecution's case.

A number of researchers have applied these alternative formulations to the challenge of characterizing climate change uncertainty and there is no final consensus on the best approach. However, so long as one carefully specifies the question to be addressed, our judgment is that all four boxes in Figure 1.1 can be appropriately handled through the use of subjective probability, allowing a wide range or a multiple set of plausible distributions to represent the high levels of uncertainty, and retaining the axioms of probability. As Smithson (1988) explains:

> One of the most frequently invoked motivations for formalisms such as possibility and Shaferian belief theory is that one number is insufficient to represent subjective belief, particularly in the face of what some writers call "ignorance"... Probabilists reply that we need not in-

---

[4] The Guidance Notes for Lead Authors of the IPCC Fourth Assessment Report (IPCC, 2005) adopted a slightly modified version of this same diagram.

[5] For reviews of these alternative formulations, see Smithson (1988) and Henrion (1999).

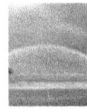

vent a new theory to handle uncertainty about probabilities. Instead we may use meta-probabilities [such as second order probability]. Even such apparently non-probabilistic concepts as possibility can be so represented…one merely induces a second-order probability distribution over the first-order subjective probabilities.

When the subjective probabilistic judgments are to be used in decision making, we believe, as outlined in Part 7, that the key issue is to employ decision criteria, such as robustness, that are appropriate to the high levels of uncertainty.

Much of the literature divides uncertainty into two broad categories, termed opaquely (for those of us who are not Latin scholars) aleatory uncertainty and epistemic uncertainty. As Paté-Cornell (1996) explains, aleatory uncertainty stems "…from variability in known (or observable) populations and, therefore, represents randomness" while epistemic uncertainty "…comes from basic lack of knowledge about fundamental phenomena (…also known in the literature as ambiguity)"[6].

While this distinction is common in much of the more theoretical literature, we believe that it is of limited utility in the context of climate and many other applied problems in assessment and decision making where most key uncertainties involve a combination of the two.

A far more useful categorization for our purposes is the split between "uncertainty about the value of empirical quantities" and "uncertainty about model functional form". The first of these may be either aleatory (the top wind speed that occurred in any Atlantic hurricane in the year 1995) or epistemic (the average global radiative forcing produced by anthropogenic aerosols at the top of the atmosphere during 1995). There is some disagreement within the community of experts on whether it is even appropriate to use the terms epistemic or aleatory when referring to a model.

Empirical quantities represent properties of the real world, which, at least in principle, can be measured. They include "…quantities in the domains of natural science and engineering, such as the oxidation rate of atmospheric pollutants, the thermal efficiency of a power plant, the failure rate of a valve, or the carcinogenic potency of a chemical, and quantities in the domain of the social sciences, such as demand elasticities or prices in economics, or judgmental biases in psychology. To be empirical, variables must be measurable, at least in principle, either now or at some time in the future.

These should be sufficiently well-specified so that they can pass the clarity test. Thus, it is permissible to express uncertainty about an empirical quantity in the form of a probability distribution. Indeed, we suggest that the only types of quantity whose uncertainty may appropriately be represented in probabilistic terms are empirical quantities[7]. This is because they are the only type of quantity that is both uncertain and can be said to have a true, as opposed to an appropriate or good value"[8].

Uncertainty about the value of an empirical quantity can arise from a variety of sources. These include lack of data; inadequate or incomplete measurement; statistical variation arising from measurement instruments and methods; systematic error and the subjective judgments needed to estimate its nature and magnitude; and inherent randomness. Uncertainty about the value of empirical quantities can also arise from sources such as the imprecise use of language in describing the quantity of interest and disagreement among different experts about how to interpret available evidence.

Not all quantities are empirical. Moreover, quantities with the same name may be empirical in some contexts and not in others. For example, quantities that represent a decision maker's own value choice or preference, such as a discount rate, coefficient of risk aversion, or the invest-

To be empirical, variables must be measurable, at least in principle, either now or at some time in the future, we suggest that the only types of quantity whose uncertainty may appropriately be represented in probabilistic terms are empirical quantities.

---

[6] The Random House Dictionary defines *aleatory* as "of or pertaining to accidental causes; of luck or chance; unpredictable" and defines *epistemic* as "of or pertaining to knowledge or the conditions for acquiring it".

---

[7] This advice is not shared by all authors. For example, Cyert and DeGroot (1987) have treated uncertainty about a decision maker's own value parameters as uncertain. But, see our discussion about in the next paragraph.

[8] Text in quotation marks in this and the preceding paragraph comes directly from the writings of two of the authors, Morgan and Henrion (1990).

ment rate to prevent mortality ("value of life") represent choices about what he or she considers to be appropriate or good. If decision makers are uncertain about what value to adopt, they should perform parametric or "switchover" analysis to explore the implications of alternative choices[9]. However, if an analyst is modeling the behavior of *other* decision makers, and needs to know how they will make such choices, then these same quantities become empirical and can appropriately be represented by a probability distribution[10].

Some authors refer to some forms of aleatory uncertainty as "variability". There are cases in which the distinction between uncertainty about the value of an empirical quantity and variability in that value (across space, time, or other relevant dimensions) is important. However, in many practical analyses, maintaining a distinction between uncertainty and variability is not especially important (Morgan and Henrion, 1990) and maintaining it can give rise to overly complicated and confusing analysis. Some people who accept only a frequentist notion of probability insist on maintaining the distinction because variability can often be described in terms of histograms or probability distributions based only on a frequentist interpretation.

A model is a simplified approximation of some underlying causal structure. Debates, such as whether a dose-response function is really linear, and whether or not it has a threshold below which no health effect occurs, are not really about what model is "true". None of these models is a complete, accurate representation of reality. The question is what is a more "useful" representation given available scientific knowledge and data and the intended use that is to be made of, or decisions to be based on, the analysis. In this sense, uncertainty about model functional form is neither aleatory nor

epistemic. The choice of model is part pragmatic. Good (1962) described such a choice of model as "type II rationality"—how can we choose a model that is a reasonable compromise between the credibility of results and the effort to create and analyze the model (collect data, estimate model parameters, apply expert judgment, compute the results, *etc.*).

Uncertainty about model functional form can arise from many of the same sources as uncertainty about the value of empirical quantities: inadequate or incomplete measurements and data that prevent the elimination of plausible alternatives; systematic errors that mislead folks in their interpretation of underlying mechanisms; inadequate imagination and inventiveness in suggesting or inferring the models that could produce the available data; and disagreement among different experts about how to interpret available evidence.

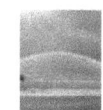

In most of the discussion that follows, by "model functional form" we will mean a description of how the world works. However, when one includes policy-analytic activities, models may also refer to considerations such as decision makers' "objectives" and the "decision rules" that they apply. These are, of course, normative choices that a decision maker or analyst must make. A fundamental problem, and potential source of uncertainty on the part of users of such analysis, is that the people who perform such analysis are often not explicit about the objectives and decision rules they are using. Indeed, sometimes they skip (unknowingly and inconsistently) from one to another decision rule in the course of doing an analysis.

It is also important to note that even when the functional form of a model is precisely known, its output may not be well known after it has run for some time. This is because some models, as well as some physical processes such as the weather and climate, are so exquisitely sensitive to initial conditions that they produce results that are chaotic (Devaney, 2003; Lorenz, 1963).

All of the preceding discussion has focused on factors and processes that we know or believe exist, but about which our knowledge is in some way incomplete. In any field such as climate

> Analysts are often
> not explicit about
> the objectives and
> decision rules
> they are using.

---

[9] In this example, a parametric analysis might ask, "What are the implications of taking the value of life to be 0.5 or 1 or 5 or 10 or 50 million dollars per death averted?" A "switchover" analysis would turn things around and ask "at what value of life" does the conclusion I read switch from Policy A to Policy B?" If the policy choice does not depend upon the choice of value across the range of interest, it may not be necessary to further refine the value.

[10] For a more detailed discussion of this and similar distinctions, see the discussion in Section 4.3 of Morgan and Henrion (1990).

change and its impacts, there are also things about which we are completely ignorant. While Donald Rumsfeld (2002) was widely lampooned in the popular press, he was absolutely correct when he noted that "...there are known unknowns. That is to say, we know there are some things we do not know. But there are also unknown unknowns, the ones we don't know we don't know".

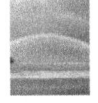

Things we know we do not know can often be addressed and sometimes understood through research. Things, about which we do not even recognize we don't know, are only revealed by adopting an always-questioning attitude toward evidence.

Things we know we do not know can often be addressed and sometimes understood through research. Things, about which we do not even recognize we don't know, are only revealed by adopting an always-questioning attitude toward evidence. This is often easier said than done. Recognizing the inconsistencies in available evidence can be difficult, since, as Thomas Kuhn (1962) has noted, we interpret the world through mental models or "paradigms" that may make it difficult to recognize and pursue important inconstancies. Weick and Sutcliffe (2001) observe that "A recurring source of misperception lies in the temptation to normalize an unexpected event in order to preserve the original expectation. The tendency to normalize is part of a larger tendency to seek confirmation for our expectations and avoid disconfirmations. This pattern ignores vast amounts of data, many of which suggest that trouble is incubating and escalating".

Freelance environmental journalist Dianne Dumanoski (quoted in Friedman *et al.*, 1999) captured this issue well when she noted:

Scientific ignorance sometimes brings many surprises. Many of the big issues we have reported on involve scientists quibbling about small degrees of uncertainty. For example, at the beginning of the debate on ozone depletion, there were arguments about whether the level or erosion of the ozone layer would be 7 percent or 13 percent within 100 years. Yet in 1985, a report came out from the British Antarctic survey, saying there was something upwards to a 50 percent loss of ozone over Antarctica. This went far beyond any scientist's worst-case scenario. Such a large loss had never been a consideration on anyone's radar screen and it certainly changed the level of the debate once it was discovered.

Uncertainty cuts both ways. In some cases, something that was considered a serious problem can turn out to be less of a threat. In other cases, something is considered less serious than it should be and we get surprised...

Perhaps the ever folksy but profound Mark Twain[11] put it best when he noted "It ain't what you don't know that gets you in trouble. It's what you know for sure that just ain't so".

---

[11] <http://www.quotedb.com/quotes/1097>.

The Importance of Quantifying Uncertainty

Lead Author: M. Granger Morgan, Carnegie Mellon Univ.

Contributing Authors: Hadi Dowlatabadi, Univ. of British Columbia; Max Henrion, Lumina Decision Systems; David Keith, Univ. of Calgary; Robert Lempert, The RAND Corporation; Sandra McBride, Duke Univ.; Mitchell Small, Carnegie Mellon Univ.; Thomas Wilbanks, Oak Ridge National Laboratory

There are a variety of words that are used to describe various degrees of uncertainty: "probable", "possible", "unlikely", "improbable", "almost impossible", *etc.* People often ask, why not simply use such words in describing uncertainty about climate change and its impacts?

Such qualitative uncertainty language is inadequate because: 1) the same words can mean very different things to different people; 2) the same words can mean very different things to the same person in different contexts; and 3) important differences in experts' judgments about mechanisms (functional relationships), and about how well key coefficients are known, can be easily masked in qualitative discussions.

Figure 2.1 illustrates the range of meaning that people attached to a set of probability words, when asked to do so in a study conducted by Wallsten *et al.* (1986), in the absence of any specific context. Mosteller and Youtz (1990) performed a review of 20 different studies of the probabili-

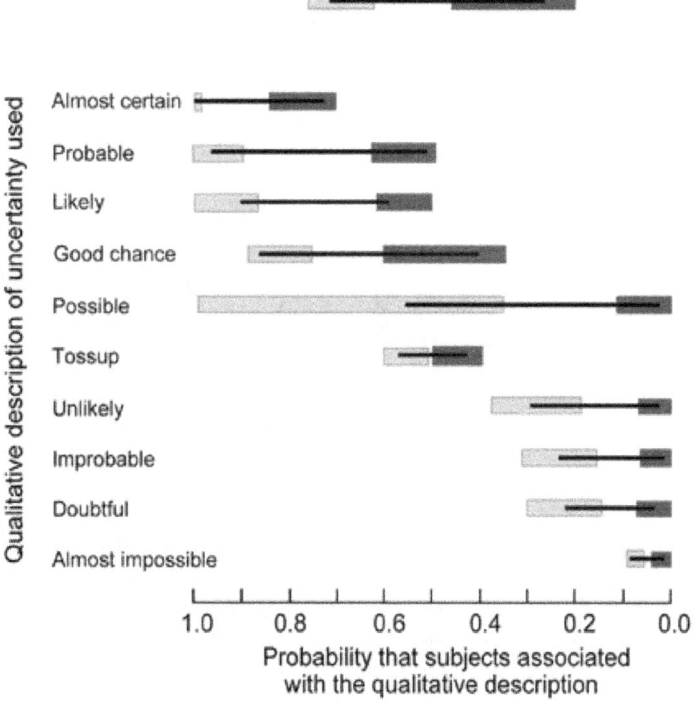

Numerical Probability Ranges for Qualitative Words

Figure 2.1 Range of numerical probabilities that respondents attached to qualitative probability words in the absence of any specific context. Figure redrawn from Wallsten *et al.* (1986)

25

Numerical Probability Assignments to Proposed Words

Key:

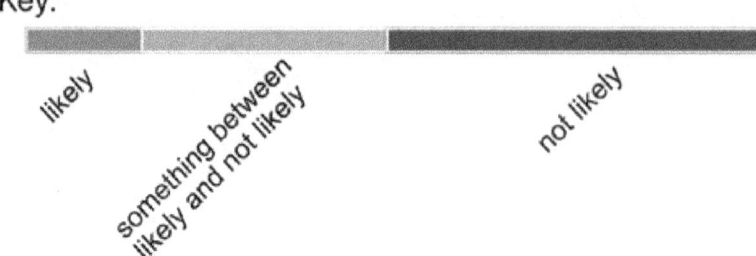

SAB members:

Other meeting participants:

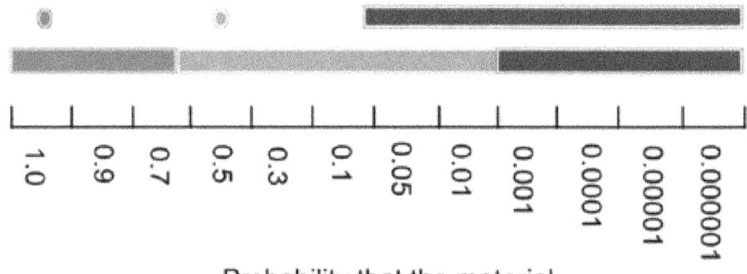

Probability that the material
is a human carcinogen

Figure 2.2 Results obtained by Morgan (1998) when members of the Executive Committee of the U.S. EPA Science Advisory Board were asked to assign numerical probabilities to words that have been proposed for use with the new U.S. EPA cancer guidelines (U.S. EPA, 1996). Note that, even in this relatively small and expert group, the minimum probability associated with the word "likely" spans four orders of magnitude, the maximum probability associated with the word "not likely" spans more than five orders of magnitude, and there is an overlap of the probabilities the different experts associated with the two words.

ties that respondents attached to 52 different qualitative expressions. They argue that "in spite of the variety of populations, format of question, instructions, and context, the variation of the averages for most of the expressions was modest..." and they suggest that it might be possible to establish a general codification that maps words into probabilities. When this paper appeared in *Statistical Science* it was accompanied by eight invited comments (Clark, 1990; Cliff, 1990; Kadane, 1990; Kruskal, 1990; Tanur, 1990; Wallsten and Budescu, 1990; Winkler, 1990; Wolf, 1990). While several commenters who have economics or statistical backgrounds commented favorably on the feasibility of a general codification based on shared natural language meaning, those with psychological backgrounds argued strongly that context and other factors make such an effort infeasible.

For example, Mosteller and Youtz argued that on the basis of their analysis of 20 studies, "likely" appears to mean 0.69 and "unlikely" means 0.16. In a study they then did in which they asked science writers to map words to probabilities, they obtained a median value for "likely" of 0.71 (interquartile range of 0.626 to 0.776) and a median value for "unlikely" of 0.172 (interquartile range of 0.098 to 0.227). In contrast, Figure 2.2 illustrates the range of numerical probabilities that individual members of the Executive Committee of the U.S. Environmental Protection Agency (EPA) Science Advisory Board attached to the words "likely" and "not likely" when those words were being used to describe the probability that a chemical agent is a human carcinogen (Morgan, 1998). Note that, even in this relatively small and expert group, the minimum probability associated with the word "likely" spans four orders of magnitude, the maximum probability associated with the word "not likely" spans more than five orders of magnitude, and there is an actual overlap of the probabilities the different experts associated with the two words! Clearly, in this setting the words do not mean roughly the same thing to all experts, and without at least some quantification, such qualitative descriptions of uncertainty convey little, if any, useful information.

While some fields, such as environmental health impact assessment, have been relatively slow to learn that it is important to be explicit about how uncertainty words are mapped into probabilities, and have resisted the use of numerical descriptions of uncertainty (Presidential/Congressional Commission on Risk Assessment and Risk Management, 1997; Morgan, 1998), the climate assessment community has made relatively good, if uneven, progress in recognizing and attempting to deal with this issue. Notable recent examples include the guidance document developed by Moss and Schneider (2000) for authors of the Intergovernmental Panel on Climate Change (IPCC) Third Assessment Report and the mapping of probability words into specific numerical values employed in the 2001 IPCC reports (IPCC, 2001a, b) (Table 2.1) and by the National Assessment Synthesis Team of the U.S. National Assessment (NAST, 2001).

Table 2.1 Mapping of probability words into quantitative subjective probability judgments, used by WG I and II of the Intergovernmental Panel on Climate Change Third Assessment (IPCC 2001a, b) based on recommendations developed by Moss and Schneider (2000).

Word	Probability range
Virtually certain	>0.99
Very likely	0.9-0.99
Likely	0.66-0.9
Medium likelihood	0.33-0.66
Unlikely	0.1-0.33
Very unlikely	0.01-0.1
Exceptionally unlikely	<0.01

Note: The report of the Intergovernmental Panel on Climate Change *Workshop on Describing Scientific Uncertainties in Climate Change to Support Analysis of Risk and of Options* (IPCC, 2004) observed: "Although WGIII TAR authors addressed uncertainties in the WG3-TAR, they did not adopt the Moss and Schneider uncertainty guidelines. The treatment of uncertainty in the WG3-AR4 can be improved over what was done in the TAR."

The climate assessment community has made relatively good, if uneven progress in recognizing and attempting to deal with how uncertainty words are mapped into probabilities.

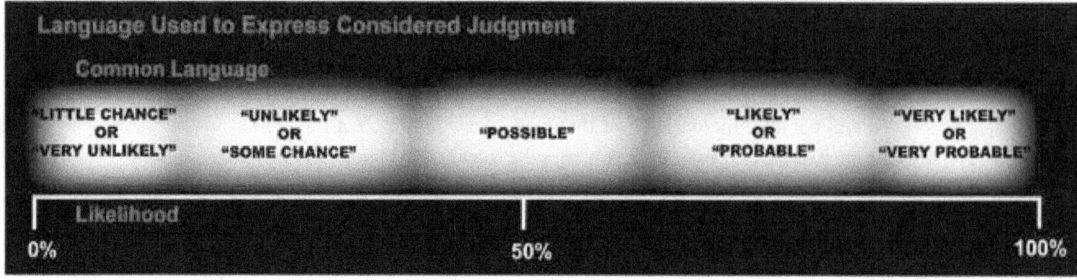

Figure 2.3 Mapping of probability words into quantitative subjective probability judgments, used in their two reports, by the members of the National Assessment Synthesis Team of the United States National Assessment (NAST, 2001).

The mapping used in the U.S. National Assessment, which the authors attempted to apply consistently throughout their two reports, is shown in Figure 2.3.

The IPCC Fourth Assessment Report drew a distinction between confidence and likelihood. They note (IPCC, 2007):

> The uncertainty guidance provided for the Fourth Assessment Report draws, for the first time, a careful distinction between levels of confidence in scientific understanding and the likelihoods of specific results. This allows authors to express high confidence that an event is extremely unlikely (*e.g.*, rolling a dice twice and getting six both times), as well as high confidence that an event is about as likely as not (*e.g.*, a tossed coin coming up heads).

The mapping used for defining levels of confidence in the Fourth Assessment is reported in Table 2.2.

Table 2.2 Mapping of probability words into quantitative subjective judgments of confidence as used in the Intergovernmental Panel on Climate Change Fourth Assessment (IPCC, 2005, 2007)

Word	Probability Range
Very High Confidence	At least 9 out of 10 chance
High Confidence	About 8 out of 10 chance
Medium Confidence	About 5 out f 10 Chance
Low Confidnece	About 2 out of 10 Chance
Very Low Confidence	Less than 1 out of 10 Chance

Note: *Guidance Notes for Lead Authors of the IPCC Fourth Assessment Report on Addressing Uncertainties (IPCC, 2005) includes both this table and Table 2.1.*

PART 3

Cognitive Challenges in Estimating Uncertainty

Lead Author: M. Granger Morgan, Carnegie Mellon Univ.

Contributing Authors: Hadi Dowlatabadi, Univ. of British Columbia; Max Henrion, Lumina Decision Systems; David Keith, Univ. of Calgary; Robert Lempert, The RAND Corporation; Sandra McBride, Duke Univ.; Mitchell Small, Carnegie Mellon Univ.; Thomas Wilbanks, Oak Ridge National Laboratory

While our brains are very good at doing many tasks, we do not come hard-wired with statistical processors. Over the past several decades, experimental psychologists have begun to identify and understand a number of the "cognitive heuristics" we use when we make judgments that involve uncertainty.

The first thing to note is that people tend to be systematically overconfident in the face of uncertainty—that is, they produce probability distributions that are much too narrow. Actual values, once they are known, often turn out to lie well outside the tails of their previous distribution. This is well illustrated with the data in the summary table reproduced in Figure 3.1. This table reports results from laboratory studies in which, using a variety of elicitation methods, subjects were asked to produce probability distributions to indicate their estimates of the value of a number of well known quantities. If the respondents were "well calibrated", then the true value of the judged quantities should fall within the 0.25 to 0.75 interval of their probability distribution about half the time. We call the frequency with which the true value actually fell within that interval the interquartile index. Similarly, the frequency with which the true value lies below the 0.01 or above the 0.99 probability values in their distribution is termed the "surprise index". Thus, for a well-calibrated respondent, the surprise index should be two percent.

In these experimental studies, interquartile indices typically were between 20 and 40 percent rather than the 50 percent they should have been, and surprise indices ranged from a low of 5 percent (2.5 times larger than it should have been) to 50 percent (25 times larger than it should have been).

Subjective Probability Distributions

	Number of Assessments N	Interquartile index (ideal 50%)	Surprise index (ideal 2%)
Alpert & Raiffa (1969)			
Group 1-A	880	33	46
Group 2 & 3	1,670	33	39
Group 4	600	36	21
Hession & McCarthy (1974) Fractiles	2,035	25	47
Selvidge (1975)			
Five fractiles	400	56	10
Seven fractiles	520	50	7
Schaefer & Borcherding (1973)			
Fractiles	396	23	39
Hypothetical sample	396	16	50
Pickhardt & Wallace (1974)	?	39	32
Group 1	?	30	46
Group 2			
Seaver, von Winterfeldt, & Edwards (1978)	160	42	34
Fractiles	160	53	24
Odds-fractiles	180	57	5
Probabilities	180	47	5
Odds	140	31	20
Log-Odds			
Stael von Holstein (1971) Fixed intervals	1,269	27	30
Murphy & Winkler (1974 & 1977)	132	45	27 (ideal 25)
Fixed intervals	432	54	21 (ideal 25)
Fractiles			
Schaefer (1976) Fixed interval	660	27	25
Lichtenstein & Fischhoff (1978) Fractiles	924	33	41
Seaver (1978) Parameters of beta dist.	3,200	29	25

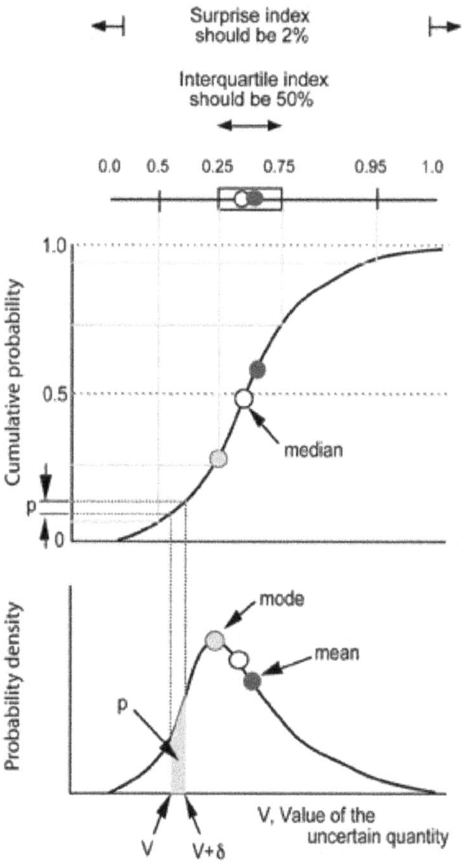

Figure 3.1 Summary of data from different studies in which, using a variety of methods, people were asked to produce probability distributions on the value of well known quantities (such as the distance between two locations), so that their distributions can be subsequently checked against true values. The results clearly demonstrate that people are systematically overconfident (*i.e.*, produce subjective probability distributions that are too narrow) when they make such judgments. The table is reproduced from Morgan and Henrion (1990) who, in compiling it, drew in part on Lichtenstein *et al.* (1982). Definitions of interquartile index and surprise index are shown in the diagram on the right.

Speed of Light: Experimental Values

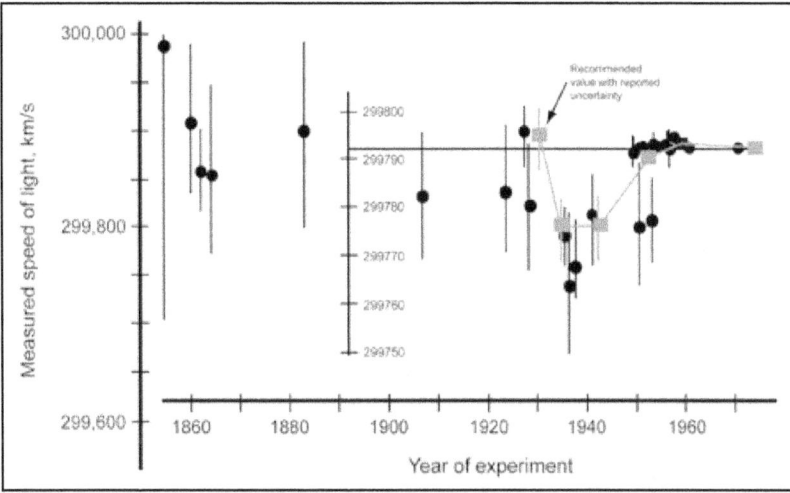

Figure 3.2 Time series of reported experimental values for the speed of light over the period
from the mid-1800's to the present (black points). Recommended values are shown in gray. These
values should include a subjective consideration of all relevant factors. Note, however, that for
a period of approximately 25 years during the early part of the last century, the uncertainty
being reported for the recommended values did not include the current best estimate. Similar
results obtained for recommended values of other basic physical quantities such as Planck's
constant, the charge and mass of the electron and Avogadro's number. For details, see Henrion
and Fischhoff (1986) from which this figure has been redrawn.

Overconfidence is not unique to non-technical judgments. Henrion and Fischhoff (1986) have examined the evolution of published estimates of a number of basic physical constants, as compared to the best modern values. Figure 3.2 shows results for the speed of light. While one might expect error bars associated with published experimental results not to include all possible sources of uncertainty, the "recommended values" do attempt to include all uncertainties. Note that for a period of approximately 25 years during the early part of the last century, the one standard deviation error bar being reported for the recommended values did not include the current best estimate.

Three cognitive heuristics are especially relevant in the context of decision making under uncertainty: availability; anchoring and adjustment; and representativeness. For a comprehensive review of much of this literature, see Kahneman *et al.* (1982).

When people judge the frequency of an uncertain event they often do so by the ease with which they can recall such events from the past, or imagine such events occurring. This "availability heuristic" serves us well in many situations. For example, if I want to judge the likeli-

hood of encountering a traffic police car on the way to the airport mid-afternoon on a work day, the ease with which I can recall such encounters from the past is probably proportional to the likelihood that I will encounter one today, since I have driven that route many times at that time of day. However, if I wanted to make the same judgment for 3:30 a.m. (a time at which I have never driven to the airport), using availability may not yield a reliable judgment.

A classic illustration of the availability heuristic in action is provided in Figure 3.3a, which shows results from a set of experimental studies conducted by Lichtenstein *et al.* (1978) in which well educated Americans were told that 50,000 people die each year in the United States from motor vehicle accidents[1], and were then asked to estimate the number of deaths that occurred each year from a number of other causes. While there is scale compression—the likelihood of high probability events is underestimated by about an order of magnitude, and the likelihood of low probability events is overestimated by a couple orders of magnitude—the fine struc-

---

[1] Today, while Americans drive more, thanks to safer cars and roads, and reduced tolerance for drunk driving, the number has fallen to about 40,000 deaths per year.

ture of the results turns out to be replicable, and clearly shows the operation of availability. Many people die of stroke, but the average American hears about such deaths only when a famous person or close relative dies, thus the probability of stroke is underestimated. Botulism poisoning is very rare, but whenever anyone dies, the event is covered extensively in the news and we all hear about it. Thus, through the operation of availability, the probability of death from botulism poisoning is overestimated. In short, judgments can be dramatically affected by what gets one's attention. Things that come readily to mind are likely to have a large effect on peoples' probabilistic judgments. Things that do not come readily to mind may be ignored. Or to paraphrase the fourteenth century proverb, all too often out of sight is out of mind.

We can also illustrate "anchoring and adjustment" with results from a similar experiment in which Lichtenstein *et al.* (1978) made no

Cognitive Heuristics of Availability and Anchoring and Adjustment

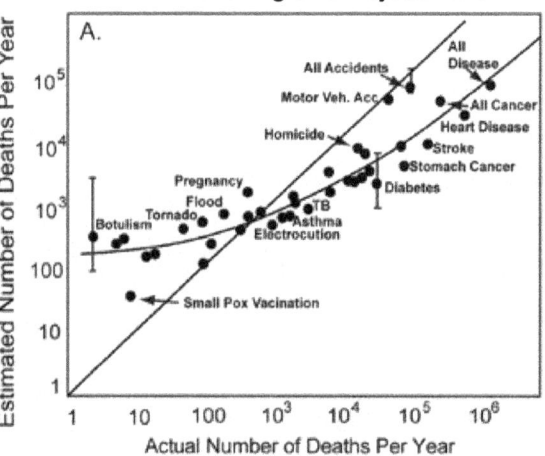

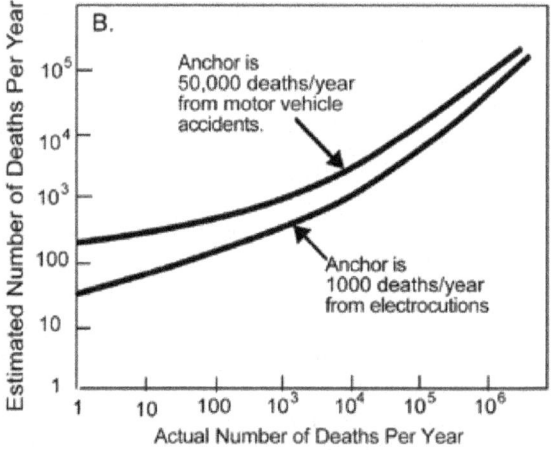

Figure 3.3 Illustration of the heuristic of availability (a) and of anchoring and adjustment (b). If respondents made perfect estimates, the results would lie along the diagonal. In the upper figure, note that stroke lies below the curved trend line and that botulism lies above the trend line—this is a result of the availability heuristic—we do not learn of most stroke deaths and we do learn of most botulism deaths via news reports. The lower figure replicates the same study with an anchor of 1,000 deaths per year. Due to the influence of this lower anchor through the heuristic of anchoring and adjustment, the mean trend line has moved down. Figures are redrawn from Lichtenstein *et al.* (1978).

mention of deaths from motor vehicle accidents but instead told a different group of respondents that about 1,000 people die each year in the United States from electrocution. Figure 3.3b shows the resulting trend lines for the two experiments. Because in this case respondents started with the much lower "anchor" (1,000 rather than 50,000), all their estimates are systematically lower.

One of the most striking experimental demonstrations of anchoring and adjustment was reported by Tversky and Kahneman (1974):

> In a demonstration of the anchoring effect, subjects were asked to estimate various quantities stated in percentages (for example, the percentage of African countries in the United Nations). For each quantity a number between 0 and 100 was determined by spinning a wheel of fortune in the subject's presence. The subjects were instructed to indicate first whether that number was higher or lower than the value of the quantity, and then to estimate the value of the quantity by moving upward or downward from the given quantity. Different groups were given different numbers for each quantity, and these arbitrary numbers had a marked effect on the estimates. For example, the median estimates of the percentage of African countries in the United Nations were 25 and 45 for groups that received 10 and 65, respectively, as starting points[2]. Payoffs for accuracy did not reduce the anchoring effect.

Very similar results are reported for similarly posed questions about other quantities such as, "What is the percentage of people in the United States today who are age 55 or older?"

The heuristic of "representativeness" says that people expect to see in single instantiations, or realizations of an event, properties that they know that a process displays in the large. Thus, for example, people judge the sequence of coin tosses HHHTTT (H = heads, T = tails) to be less likely than the sequence HTHHTH because the former looks less random than the latter, and

they know that the process of tossing a fair coin is a random process.

Psychologists refer to feeling and emotion as "affect". Slovic *et al.* (2004) suggest that:

> Perhaps the biases in probability and frequency judgment that have been attributed to the availability heuristic... may be due, at least in part, to affect. Availability may work not only through ease of recall or imaginability, but because remembered and imagined images come tagged with affect.

Slovic *et al.* (2004) argue that there are two fundamental ways that people make judgments about risk and uncertainty—one, the "analytic system", and the other, the "experiential system". They note that while the analytic system "...is rather slow, effortful and requires conscious control", the experiential system is "intuitive, fast, mostly automatic, and not very accessible to conscious awareness". They note that both are subject to various biases and argue both are often needed for good decision making:

> Even such prototypical analytic exercises as proving a mathematical theorem or selecting a move in chess benefit from experiential guidance, the mathematician senses whether the proof "looks good" and the chess master gauges whether a contemplated move "feels right", based upon stored knowledge of a large number of winning patterns (DeGroot, 1965 as paraphrased by Slovic *et al.*, 2004).

Psychologists working in the general area of risk and decision making under uncertainty are somewhat divided about the role played by emotions and feelings (*i.e.*, affect) in making risk and related judgments. Some (*e.g.*, Sjöberg, 2006) argue that such influences are minor; others (*e.g.*, Loewenstein, 1996; Loewenstein *et al.*, 2001) assign them a dominant role. Agreeing with Slovic *et al.*'s conclusion that both are often important, Wardman (2006) suggests that the most effective responses "...may in fact occur when they are driven by both affective and deliberative-analytical considerations, and that it is the absence of one or the other that may cause problems..."

---

[2]   Hastie and Dawes (2001) report that at the time the experiment was conducted the actual value was 35 percent.

PART 4

Statistical Methods and Models

Lead Author: M. Granger Morgan, Carnegie Mellon Univ.

Contributing Authors: Hadi Dowlatabadi, Univ. of British Columbia; Max Henrion, Lumina Decision Systems; David Keith, Univ. of Calgary; Robert Lempert, The RAND Corporation; Sandra McBride, Duke Univ.; Mitchell Small, Carnegie Mellon Univ.; Thomas Wilbanks, Oak Ridge National Laboratory

Statistical methods and models play a key role in the interpretation and synthesis of observed climate data and the predictions of numerical climate models. Important advances have been made in the development and application of both frequentist and Bayesian statistical approaches and, as noted previously, the methods yield similar results when either an uninformed prior is used for the Bayesian analysis or a very large dataset is available for estimation. Recent reviews of statistical methods for climate assessment are summarized, including procedures for trend detection, assessing model fit, downscaling, and data-model assimilation. Methods for hypothesis testing and model selection are presented, and emerging issues in statistical methods development are considered.

Levine and Berliner (1999) review statistical methods for detecting and attributing climate change signals in the face of high natural variations in the weather and climate, focusing on "fingerprint" methods designed to maximize the signal-to-noise ratio in an observed climatic dataset (Hasselmann, 1979, 1993). The climate change detection problem is framed in terms of statistical hypothesis testing and the fingerprint method is shown to be analogous to stepwise regression of the observed data (*e.g.*, temperature) against the hypothesized input signals (carbon dioxide concentrations, aerosols, *etc.*). Explanatory variables are added to the model until their coefficients are no longer statistically significant. The formulation and interpretation of the hypothesis test is complicated considerably by the complex spatial and temporal correlation structure of the dependent and explanatory variables, and Levine and Berliner discuss various approaches for addressing these concerns. The selection of the best filter for isolating a climate change signal within the natural climate record is shown to be equivalent to the determination of an optimal (most powerful) statistical test of hypothesis.

Solow (2003) reviews various statistical models used in atmospheric and climate science, including methods for:

- fitting multivariate spatial-time series models, using methods such as principal component analysis (PCA) to consider spatial covariance, and predictive oscillation patterns (PROPS) analysis and maximum covariance analysis (MCA) for addressing both spatial and temporal variations (Kooperberg and O'Sullivan, 1996; Salim et al., 2005);

- identifying trends in the rate of occurrence of extreme events given only a partially observed historical record (Solow and Moore, 2000, 2002);

- downscaling General Circulation Model (GCM) predictions to estimate climate variables at finer temporal and spatial resolution (Berliner et al., 1999; Berliner, 2003);

- assessing the goodness-of-fit of GCMs to observed data (McAvaney et al., 2001), where goodness-of-fit is often measured by the ability of the model to reproduce the observed climate variability (Levine and Berliner, 1999; Bell et al., 2000); and

- data assimilation methods that combine model projections with the observed data for improved overall prediction (Daley, 1997), including multi-model assimilation methods (Stephenson et al., 2005) and extended Kalman filter procedures that also provide for model parameter estimation (Evensen and van Leeuwen, 2000; Annan, 2005; Annan et al., 2005).

Zwiers and von Storch (2004) also review the role of statistics in climate research, focusing on statistical methods for identifying the dynamics of the climate system and implications for data collection, forecasting, and climate change detection. The authors argue that empirical models for the spatiotemporal features of the climate record should be associated with plausible physical models and interpretations for the system dynamics. Statistical assessments of data homogeneity are noted as essential when evaluating long-term records where measurement methods, local processes, and other non-climate influences are liable to result in gradual or abrupt changes in the data record (Vincent, 1998; Lund and Reeves, 2002).

Statistical procedures are reviewed for assessing the potential predictability and accuracy of future weather and climate forecasts, including those based on the data-model assimilation methods described above. Zwiers and Storch offer that for the critical tasks of determining the inherent (irreducible) uncertainty in climate predictions *versus* the potential value of learning from better data and models, Bayesian statistical methods are often better suited than are frequentist approaches.

Methods for Hypothesis and Model Testing

A well-established measure in classical statistics for comparing competing models (or hypotheses) is the likelihood ratio (LR), which follows from the common use of the maximum likelihood estimate for parameter estimation. For two competing models, M_1 and M_2, the LR is the ratio of the likelihood or maximum probability of the observed data under M_1 divided by the likelihood of the observed data under M_2, with large values of the likelihood ratio indicating support for M_1. Solow and Moore (2000) applied the LR test to look for evidence of a trend in a partially incomplete hurricane record, using a Poisson distribution for the number of hurricanes in a year with a constant sighting probability over the incomplete record period. The existence of such a trend could indicate warming in the North Atlantic Basin, but based on their analysis, little evidence was apparent. In cases such as that above in which the LR tests models with the same parameterization and simple hypotheses are of interest, the LR is equivalent to the Bayes Factor, which is the ratio of the posterior odds of M_1 to the prior odds of M_1. That is, the Bayes Factor represents the odds of favoring M_1 over M_2 based solely on the data, and thus the magnitude of the Bayes Factor is often used as a measure of evidence in favor of M_1.

An approximation to the log of the Bayes Factor for large sample sizes, Schwarz's Bayesian Information Criterion or BIC, is often used as a model-fitting criterion when selecting among all possible subset models. The BIC allows models to be evaluated in terms of a lack of fit component (a function of the sample size and mean squared error) and a penalty term for the number of parameters in a model. The BIC

A well-established measure in classical statistics for comparing competing models (or hypotheses) is the liklihood ratio (LR). For two competing models, the LR is the ratio of the liklihood or maximum probability of the observed data under the first model divided by the liklihood of the observed data under the second model, with large values of the likelihood ratio indicating support for the first model.

differs from the well-known Akaike's Information Criterion (AIC) only in the penalty for the number of included model terms. Another related model selection statistic is Mallow's Cp (Laud and Ibrahim, 1995). Karl *et al.* (1996) utilize the BIC to select among autoregressive moving average (ARMA) models for climate change, finding that the Climate Extremes Index (CEI) and the United States Greenhouse Climate Response Index (GCRI) increased abruptly during the 1970s.

Model uncertainty can also be addressed by aggregating the results of competing models into a single analysis. For instance, in Part 5 we report an estimate of climate sensitivity (Andronova and Schlesinger, 2001) made by simulating the observed hemispheric-mean near-surface temperature changes since 1856 with a simple climate/ocean model forced radiatively by greenhouse gases, sulfate aerosols, and solar-irradiance variations. A number of other investigators have used models together with historical climate data and other evidence to develop probability distributions for climate sensitivity or bound estimates of climate sensitivity or other variables. Several additional efforts of this sort are discussed in Part 5. An increasing number of these studies have begun to employ Bayesian statistical methods (*e.g.,* Epstein, 1985; Berliner *et al.,* 2000; Katz, 2002; Tebaldi *et al.,* 2004, 2005).

As noted in Katz (2002) and Goldstein (2006), Bayesian methods bring a number of conceptual and computational advantages when characterizing uncertainty for complex systems such as those encountered in climate assessment. Bayesian methods are particularly well suited for problems where experts differ in their scientific assessment of critical processes and parameter values in ways that cannot, as yet, be resolved by the observational record. Comparisons across experts not only help to characterize current uncertainty, but help to identify the type and amount of further data collection likely to lead to resolution of these differences. Bayesian methods also adapt well to situations where hierarchical modeling is needed, such as where model parameters for particular regions, locations, or times can be viewed as being sampled from a more general (*e.g.,* global) distribution of parameter values (Wilke *et al.,* 1998).

Bayesian methods are also used for uncertainty analysis of large computational models, where statistical models that emulate the complex, multidimensional model input-output relationship are learned and updated as more numerical experiments are conducted (Kennedy and O'Hagan, 2001; Fuentes *et al.,* 2003; Kennedy *et al.,* 2006; Goldstein and Rougier, 2006). In addition, Bayesian formulations allow the predictions from multiple models to be averaged or weighted in accordance with their consistency with the historical climate data (Wintle *et al.,* 2003; Tebaldi *et al.,* 2004, 2005; Raftery *et al.,* 2005; Katz and Ehrendorfer, 2006; Min and Hense, 2006).

Regardless of whether frequentist or Bayesian statistical methods are used, the presence of uncertainty in model parameters and the models themselves calls for extensive sensitivity analysis of results to model assumptions. In the Bayesian context, Berger (1994) reviews developments in the study of the sensitivity of Bayesian answers to uncertain inputs, known as robust Bayesian analysis. Results from Bayesian modeling with informed priors should be compared to results generated from priors incorporating more uncertainty, such as flat-tailed distributions, non-informative and partially informative priors. Sensitivity analysis on the likelihood function and the prior by consideration of both non-parametric and parametric classes is often called for when experts differ in their interpretation of an experiment or a measured indicator. For example, Berliner *et al.* (2000) employ Bayesian robustness techniques in the context of a Bayesian fingerprinting methodology for assessment of anthropogenic impacts on climate by examining the range of posterior inference as prior inputs are varied. Of note, Berliner *et al.* also compare their results to those from a classical hypothesis testing approach, emphasizing the conservatism of the Bayesian method that results through more attention to the broader role and impact of uncertainty.

Comparisons across experts not only help to characterize current uncertainty, but help to identify the type and amount of further data collection likely to lead to resolution of scientific assessment of critical processes and parameter values in ways that cannot, as yet, be resolved by the observational record.

Emerging Methods and Applications

While the suite of tools for statistical evaluation of climate data and models has grown considerably in the last two decades, new applications, hypotheses, and datasets continue to expand the need for new approaches. For example, more sophisticated tests of hypothesis can be made by testing probability distributions for uncertain parameters, rather than single nominal values (Kheshgi and White, 2001). While much of the methods development to date has focused on atmospheric-oceanic applications, statistical methods are also being developed to address the special features of downstream datasets, such as stream flow (Allen and Ingram, 2002; Koutsoyiannis, 2003; Kallache *et al.*, 2005) and species abundance (Austin, 2002; Parmesan and Yohe, 2003).

As models become increasingly sophisticated, requiring more spatial and temporal inputs and parameters, new methods will be needed to allow our limited datasets to keep up with the requirements of these models. Two recent examples are of note. Edwards and Marsh (2005) present a "simplified climate model" with a "fully 3-D, frictional geostrophic ocean component, an Energy and Moisture Balance atmosphere, and a dynamic and thermodynamic sea-ice model...representing a first attempt at tuning a 3-D climate model by a strictly defined procedure". While estimates of overturning and ocean heat transport are "well reproduced", "model parameters were only weakly constrained by the data". Jones *et al.* (2006) present an integrated climate-carbon cycle model to assess the implications of carbon cycle feedback considering parameter and model structure uncertainty. While the authors find that the observational record significantly constrains permissible emissions, the observed data (in this case also) "proves to be insufficient to tightly constrain carbon cycle processes or future feedback strength with implication for climate-carbon cycle model evaluation". Improved data collection, modeling capabilities, and statistical methods must clearly all be developed concomitantly to allow uncertainties to be addressed effectively.

> While the suite of tools for statistical evaluation of climate data and models has grown considerably in the last two decades, new applications, hypotheses, and datasets continue to expand the need for new approaches.

Box 4.1: Predicting Rainfall: An Illustration of Frequentist and Bayesian Approaches

Consider how we use probability theory in weather prediction. We have a vast storehouse of observations of temperature, humidity, cloud cover, wind speed and direction, and atmospheric pressure for a given location. These allow the construction of a classic or frequentist table of probabilities showing the observed probability of rainfall, given particular conditions. This underscores the fact that observations of a stable system permit the construction of powerful predictive models, even if underlying physical processes are not known fully.

So long as the same underlying conditions prevail, the predictive model based on historical weather will remain powerful. However, if an underlying factor does change, the predictive power of the model will fall and the missing explanatory variables will have to be discovered. More advanced stochastic models for precipitation have been developed in recent years, conditioning rainfall occurrence and amounts on atmospheric circulation patterns (e.g., Hughes et al. 1999; Charles et al., 2004). If climate-induced changes in atmospheric circulation can be predicted, projections of the statistical properties of associated precipitation fields can then be derived. As another example of uncertainty induced by changing conditions, reduced air pollution could in some locations cause the concentration of cloud condensation nuclei (CCN) to decline, affecting cloud stability and droplet formation. Under these conditions it would be useful to consider a Bayesian approach in which CCN are considered a potential additional explanatory variable. We can start with the previous model of precipitation occurrence, then modify its probability of rainfall, given different concentrations of CCN. With each observation of atmospheric aerosols and precipitation, our prior estimates of the rainfall-CCN relationship and overall rainfall occurrence will be modified eventually leading to a new more powerful model, this time inclusive of the new explanatory variable.

Ideally, we want the full distribution of rainfall in a location. This has proven difficult to do, using the frequentist method, especially when we focus on high impact events such as extreme droughts and floods. These occur too infrequently for us to use a large body of observations so we must "assume" a probability distribution for such events in order to predict their probability of occurrence. While it may be informed by basic science, there is no objective method defining the appropriate probability distribution function. What we choose to use is subjective.

Furthermore, the determinants of rainfall have been more numerous than once believed, often varying dramatically even on a decadal scale. For example, in the mid twentieth century, it was thought possible to characterize the rainfall in any location from 30 years of observations. This approach used the meteorological data for the period: 1931 to 1960 to define the climate norm around the earth. By the mid-80s however, it was clear that that 30-year period did not provide an adequate basis for predicting rainfall in the subsequent years. In short, we learned that there is no "representative" sample of data in the classical sense. What we have is an evolving condition where tele-connections such as El Niño Southern Oscillation (ENSO) and the North Atlantic Oscillation (NAO), as well as air pollution and other factors determine cloud formation, stability and rainfall.

As we gain experience with the complex of processes leading to precipitation, we also develop a sense of humility about the incomplete state of our knowledge. This is where the subjectivity in Bayesian statistics comes to the fore. It states explicitly that our predictions are contingent on our current state of knowledge and that knowledge will be evolving with new observations.

Methods for Estimating Uncertainty

Lead Author: M. Granger Morgan, Carnegie Mellon Univ.

Contributing Authors: Hadi Dowlatabadi, Univ. of British Columbia; Max Henrion, Lumina Decision Systems; David Keith, Univ. of Calgary; Robert Lempert, The RAND Corporation; Sandra McBride, Duke Univ.; Mitchell Small, Carnegie Mellon Univ.; Thomas Wilbanks, Oak Ridge National Laboratory

Many of the key variables and functional relationships that are important to understanding the climate system and how the climate may change over the coming decades and centuries will likely remain uncertain for years to come. While a variety of evidence can be brought to bear to gain insight about these uncertainties, in most cases no single piece of evidence or experimental result can provide definitive answers. Yet research planners, groups attempting to do impact assessment, policy makers addressing emissions reductions, public and private parties making long-lived capital investment decisions, and many others, all need some informed judgment about the nature and extent of the associated uncertainties.

Model-Generated Uncertainty Estimates

In some cases, probability distributions for key climate parameters can be extracted directly from available data and models. Note, however, that the models themselves often contain a myriad of implicit expert judgments. In recent years, several research groups have derived probability distributions for climate sensitivity via statistical comparisons of climate model results to recent climate records. For instance, Figure 5.1 shows an estimate of climate sensitivity (Andronova and Schlesinger, 2001) made by simulating the observed hemispheric-mean near-surface temperature changes since 1856 with a simple climate/ocean model forced radiatively by greenhouse gases, sulfate aerosols, and solar-irradiance

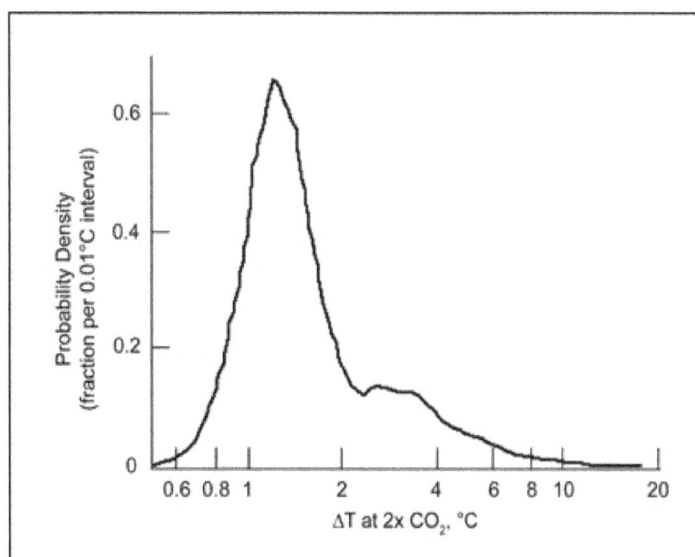

Figure 5.1 The probability density function for climate sensitivity (ΔT at 2x) estimated by Andronova and Schlesinger (2001). Using coupled atmosphere-ocean models, the observed near-surface temperature record and a bootstrap re-sampling technique, the authors examined the effect of natural variability and uncertainty in climatic radiative forcing on estimates of temperature change from the mid-nineteenth century to the present. (Figure redrawn from Andronova and Schlesinger, 2001.)

variations. The authors account for uncertainty in climatic radiative forcing by considering 16 radiative forcing models. To account for natural variability in instrumental measurements of temperature, a bootstrap procedure is used to generate surrogate observed temperature records. Figure 5.1 shows the probability distribution function for estimated climate sensitivity based on 80,000 model runs, aggregated across radiative forcing models and bootstrapped temperature records. The resultant 90-percent confidence interval for temperature sensitivity is between 1.0°C and 9.2°C. A number of other investigators have also used models together with historical climate data and other evidence to develop probability distributions for climate sensitivity or bound estimates of climate sensitivity or other variables. Several additional efforts of this sort are discussed in Part 6.

Researchers have also used data and models to derive uncertainty estimates for future socio-economic and technological driving forces. For instance, Gritsevskyi and Nakićenović (2000) and Nakićenović and Riahi (2002) have estimated probability distributions for the investment costs and learning rates of new technologies based on the historical distributions of cost and performance for many similar technologies and then used these probability estimates to forecast distributions of future emission paths. Some authors have estimated probability distributions for future emissions by assessing the frequency of results over different emissions models or by propagating subjective probability distributions for key inputs through such emission models (Webster *et al.*, 2003). Such approaches can suggest which uncertainties are most important in determining any significant deviations from a base-case projection and can prove particularly important in helping to make clear when proposed emissions scenarios differ in important ways from past trends. Care must be taken, however, with such estimates because unlike physical parameters of the climate system, socioeconomic and technological factors need not

Expert Elicitation on Climate Sensitivity

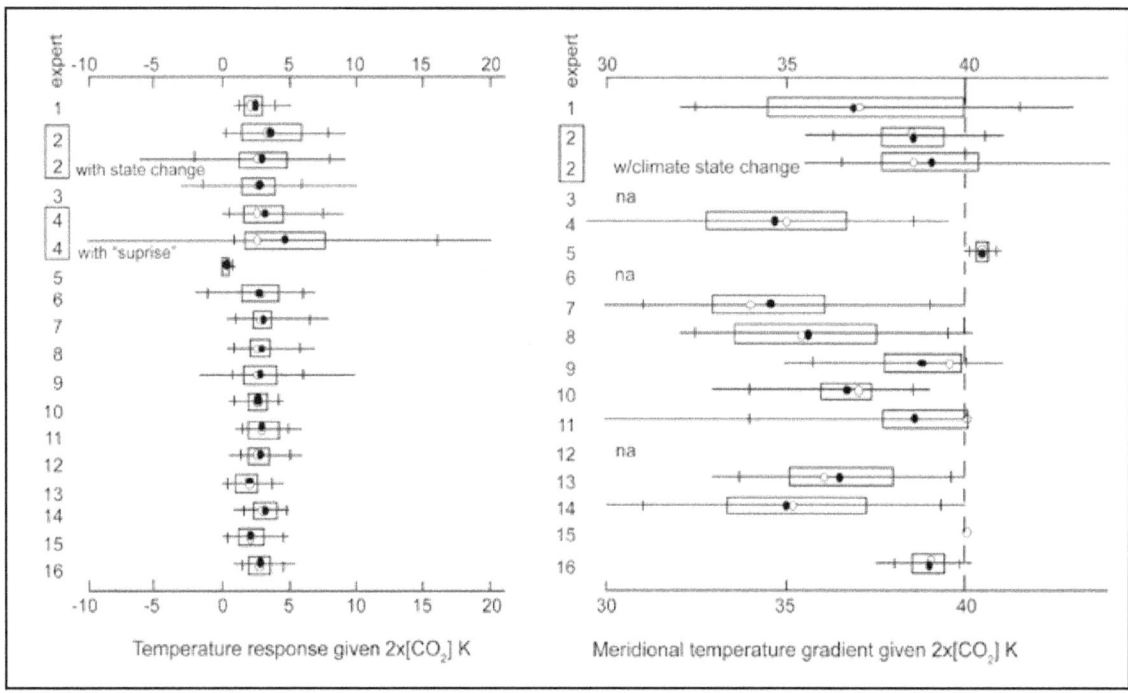

Figure 5.2 Examples of results from expert elicitations conducted by Morgan and Keith (1995) reported as box plots. Climate sensitivity is shown on the left and pole-to-equator temperature gradient on the right. Horizontal lines in the box plots report the full range of the distribution; vertical tick marks show the 0.95 confidence intervals; boxes report the 0.25 to 0.75 central interval; open dots are best estimates and closed dots are means of the distributions. While there is apparently large agreement among all but one of the experts about the climate sensitivity, a quantity that has been widely discussed, judgments about the closely related pole-to-equator temperature gradient show much greater inter-expert variability and even some disagreement about the sign of the change from the current value which is indicated by the vertical dashed line.

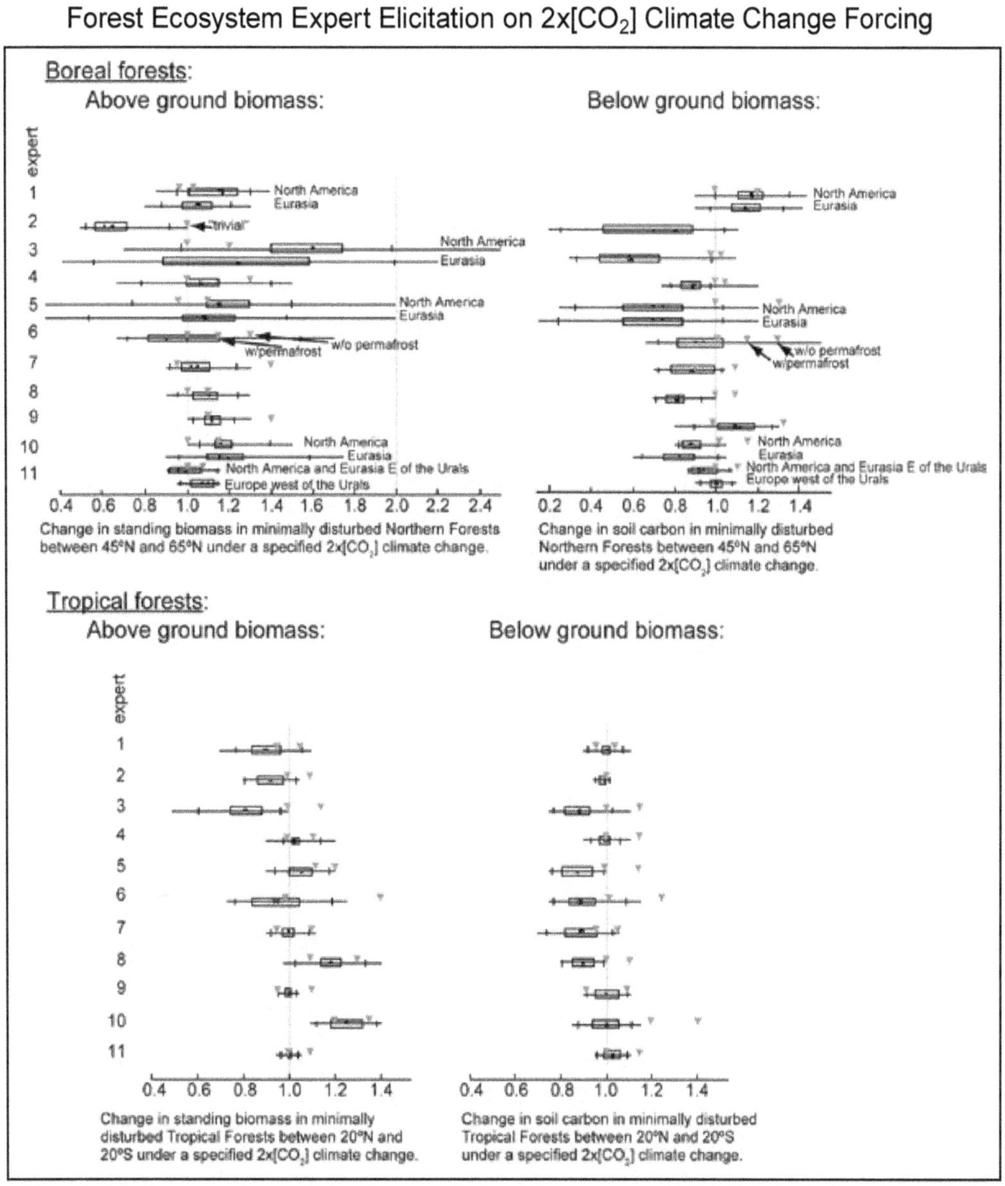

Forest Ecosystem Expert Elicitation on 2x[CO$_2$] Climate Change Forcing

Figure 5.3 Examples of results from expert elicitations of forest ecosystem experts on change in above- and below-ground biomass for a specified 2x[CO$_2$] climate change forcing (Morgan *et al.*, 2001). Note that in several cases there is not even agreement about the sign of the impact on carbon stocks. Notation is the same as in Figure 5.2. Gray inverted triangles show ranges for changes due to doubling of atmospheric carbon dioxide (CO$_2$), excluding a climate effect

remain constant over time and may be strongly interrelated and conditional on each other. Since we expect the twenty-first century will differ in important ways from the twentieth, as the twentieth differed in important ways from the nineteenth, *etc.*, we should regard these uncertainty estimates of future socio-economic outcomes with less confidence than those of physical parameters of the climate system when they are thought to be fundamentally constant through time.

Expert Elicitation

Model and data generated uncertainty estimates can be very valuable in many cases. In particular, they are most germane for judgments about well-established knowledge, represented by the upper right-hand corner of Figure 1.1[1]. But in many situations, limitations of data, scientific understanding, and the predictive capacity of

_____
[1] The drive to produce estimates using model-based methods may also stem from a reluctance to confront the use of expert judgment explicitly.

Expert Elicitation on Aerosol Forcing

When model and
data generated
uncertainty estimates
are unavailable, the
best strategy is to
ask a number of
leading experts to
consider and carefully
synthesize the full
range of current
scientific theory and
available evidence
and then provide
their judgments
in the form of
subjective probability
distributions.

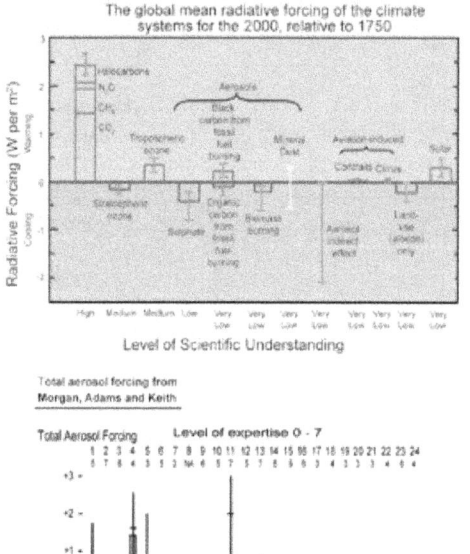

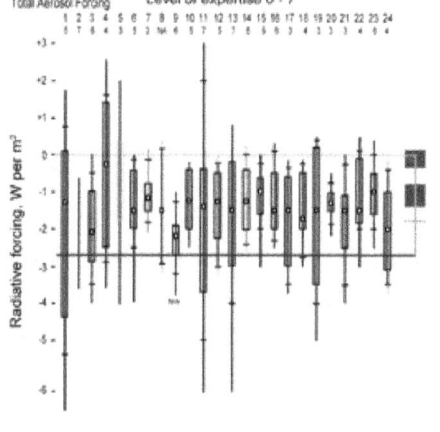

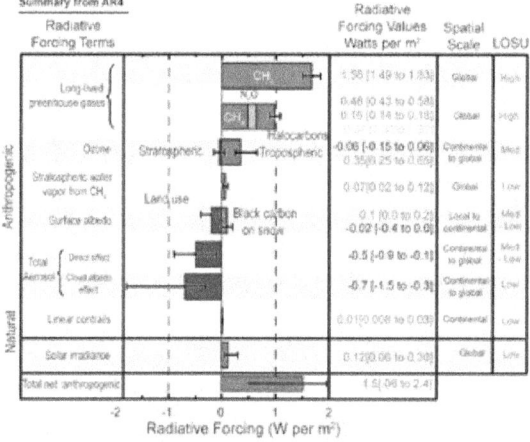

Figure 5.4 Comparison of estimates of aerosol forcing from the Intergovernmental Panel on Climate Change (IPCC) Third Assessment (TAR) (left), an expert elicitation of 24 leading aerosol experts (Morgan *et al.*, 2006) (center) and the IPCC Fourth Assessment Report (AR4) (right). All radiative forcing scales (in watts per square meter [W per m²]) are identical. Uncertainty ranges in the AR4 are 90-percent confidence intervals. The horizontal tick marks on the box plots in center are also 90-percent confidence intervals. Note that even if one simply adds the 90-percent outer confidence interval for the two AR4 estimates (a procedure that overstates the overall uncertainty in the AR4 summary), 13 of the 24 experts (54 percent) interviewed produced lower 5-percent confidence value that lie below that line, and 7 out of 24 (29 percent) produced upper 5-percent confidence values above upper bound from AR4. This comparison suggests that the uncertainty estimates of aerosol forcing reported in AR4 are tighter than those of many individual experts who were working in the field at about the same time as the AR4 summary was produced.

models will make such estimates unavailable, with the result that they must be supplemented with other sources of information.

In such circumstances, the best strategy is to ask a number of leading experts to consider and carefully synthesize the full range of current scientific theory and available evidence and then provide their judgments in the form of subjective probability distributions.

Such formal individually-focused elicitation of expert judgment has been widely used in applied Bayesian decision analysis (DeGroot, 1970; Spetzler and Staël von Holstein, 1975; Watson and Buede, 1987; von Winterfeldt and Edwards, 1986; Morgan and Henrion, 1990; Cooke, 1991), often in business applications, and in climate and other areas of environmental policy through the process of "expert elicitation" (Morgan *et al.*, 1978a, 1978b, 1984, 1985, 2001, 2006; National Defense University, 1978; Wallsten and Whitfield, 1986; Stewart *et al.*, 1992; Nordhaus, 1994; Evans *et al.*, 1994a, b; Morgan and Keith, 1995; Budnitz *et al.*,1995, 1998; Garthwaite *et al.*, 2005). An advantage of such expert elicitation is that it can effectively enumerate the range of expert judgments unhampered by social interactions, which may constrain discussion of extreme views in group-based settings.

Figures 5.2, 5.3, and 5.4 provide examples of results from expert elicitations done respectively on climate science in 1995, on forest ecosystem impacts in 2001, and on aerosol forcing in 2005. These are summary plots. Much greater detail, including judgments of time dynamics, and research needs are available in the relevant papers.

The comparison of individual expert judgments in Figure 5.4 with the summary judgment of the Intergovernmental Panel on Climate Change (IPCC) Fourth Assessment Report (IPCC, 2007) suggests that the IPCC estimate of uncertainty in total aerosol forcing may be overconfident. Similar results are apparent when comparing Zickfeld *et al.* (2007) with IPCC estimates related to the Atlantic meridional overturning circulation (AMOC). Indeed, the *Guidance Notes for Lead Authors of the IPCC Fourth Assessment Report on Address-*

ing Uncertainties (IPCC, 2005) warns authors against this tendency:

> Be aware of the tendency of a group to converge on an expressed value and become overconfident in it. Views and estimates can also become anchored on previous versions or values to a greater extent than is justified. Recognize when individual views are adjusting as a result of group interactions and allow adequate time for such changes in viewpoint to be resolved.

In light of what they see as insufficient success in overcoming these and other problems, Oppenheimer *et al.* (2007) have suggested that current strategies for producing IPCC summary statements of uncertainty need to be reassessed.

Of course, expert judgment is not a substitute for definitive scientific research. Nor is it a substitute for careful deliberative expert reviews of the literature of the sort undertaken by the IPCC. However, its use within such review processes could enable a better expression of the diversity of expert judgment and allow more formal expression of expert judgments, which are not adequately reflected, in the existing literature. It can also provide insights for policy makers and research planners while research to produce more definitive results is ongoing. It is for these reasons that Moss and Schneider (2000) have argued that such elicitations should become a standard input to the IPCC assessment process.

In selecting experts to participate in an expert elicitation, it is important to draw upon representatives from across all the relevant disciplines and schools of thought. At the same time, this process is fundamentally different from that of drawing a random sample to estimate some underlying true value. In the case of expert elicitation, it is entirely possible that one expert, perhaps even one whose views are an outlier, may be correctly reflecting the underlying physical reality, and all the others may be wrong. For this same reason, when different experts hold different views it is often best not to combine the results before using them in analysis, but rather to explore the implications of each expert's views so that decision makers

In selecting experts to participate in an expert elicitation, it is important to draw upon representatives from across all the relevant disciplines and schools of thought.

have a clear understanding of whether and how much the differences matter in the context of the overall decision (Morgan and Henrion, 1990; Keith, 1996).

It has been our experience that when asked to participate in such elicitation exercises, with very few exceptions, experts strive to provide their best judgments about the quantity or issue at hand, without considering how those judgments might be used or the implications they may carry for the conclusions that may be drawn when they are subsequently incorporated in models or other analysis. In addition to the strong sense of professional integrity possessed by most leading experts, the risk of possible "motivational bias" in experts' responses in elicitation processes is further reduced by the fact that even if the results are nominally anonymous, respondents know that they may be called upon to defend their responses to their peers[2].

As noted in Part 2, unless they are accompanied by some form of quantitative calibration, qualitative summaries of uncertainty can often mask large disagreements, since the same descriptors of qualitative uncertainty can mean very different things to different people. Thus, a quantitative expert elicitation can often provide a better indication of the diversity of opinion within an expert community than is provided in many consensus summaries. For example, the expert elicitation of climate change damage estimates by Nordhaus (1994) revealed a systematic divide between social and natural scientists' considered opinions. Such results can allow others to draw their own conclusions about how important the range of expert opinions is to the overall policy debate. Sometimes apparent deep disagreements make little difference to the policy conclusions; sometimes they are of critical importance (Morgan *et al.*, 1984; Morgan and Henrion, 1990).

We believe that in most cases it is best to avoid discussion of second-order uncertainty. Very often people are interested in using ranges or even second-order probability distributions

> With very few exceptions, experts strive to provide their best judgments about the quantity or issue at hand, without considering how those judgments might be used or the implications they may carry for the conclusions that may be drawn when they are subsequently incorporated in models or other analysis.

on probabilities—to express "uncertainty about their uncertainty". In our experience, this usually arises from an implicit confusion that there is a "true" probability out there, in the same way that there is a true value for the rainfall in a specific location last year—and people want to express uncertainty about that "true" probability. Of course, there is no such thing. The probability itself is a way to express uncertainty. A second-order distribution rarely adds anything useful.

It is, of course, possible to use a second-order distribution to express the possible effect of specific new information on a probability. For example, suppose your probability that there will be an increase of more than 1°C in average global temperature by 2020 is 0.5. It makes sense then to ask, "What is your current probability distribution over the probability you will assess for that event in five years time, when you will have seen five years more climate data and climate research?" Bayesians sometimes call this a pre-posterior distribution. Note that the pre-posterior distribution is a representation of the informativeness of a defined but currently unknown source of information, in this case the next five years of data. It depends specifically on your beliefs about that information source.

Most people find pre-posterior distributions hard to think about. It is possible to use them in elicitations (Morgan and Keith, 1995). However, in public forums, they are often confused with ambiguity and other kinds of second-order probability and are liable to provoke ideological debates with proponents of alternative formalisms of uncertainty. Hence, our view is that it is usually wisest to avoid them in public forums and reserve them for that subset of specialist applications where they are really needed. This is particularly true when one is already eliciting full probability distributions about the value of uncertain quantities.

There is one exception to this general guidance, which perhaps deserves special treatment. Suppose we have two experts, A and B, who are both asked to judge the probability that a well-specified event will occur (*i.e.*, not a full probability density function [PDF] but just a single probability on the binary yes/no outcome). Suppose A knows a great deal about the

---
[2] Despite these factors, retaining consistency with prior public positions, or influence from funding sources or political beliefs is always possible, as it is in most human endeavors.

relevant science and B knows relatively little, but they both judge the probability of the event's occurrence to be 0.3. In this case, A might give a rather tight distribution if asked to state how confident he is about his judgment (or how likely he thinks it is that additional information would modify that judgment) while B should give a rather broad distribution. In this case, the resulting distribution provides a way for the two experts to provide information about the confidence they have in their judgment.

To date, elicitation of individual experts has been the most widely used method of using expert judgment to characterize uncertainty about climate-related issues. After experts have provided their responses, many of these studies later give participants the opportunity to review their own results and those of others, and make revisions should they so desire, but they are not focused on trying to achieve group consensus.

While they have not seen extensive use in climate applications, there are a number of group-based methods that have been used in other settings. Of these, the best known is the Delphi method (Dalkey, 1969; Dalkey *et al.*, 1970; Linstone and Turoff, 1975). Delphi studies involve multiple rounds in which participants are asked to make and explain judgments about uncertain quantities of interest, and then are iteratively shown the judgments and explanations of others, and asked to make revisions, in the hope that over time a consensus judgment will emerge. Such a procedure typically will not support the depth of technical detail that has been characteristic of some of the protocols that have been used in elicitation of individual climate experts.

The Delphi method was originally developed in the 1960s and was widely used as a way to combine expert judgment (Dalkey, 1969; Dalkey *et al.*, 1970; Linstone and Turoff, 1975). It involves experts iteratively making probabilistic judgments after reviewing the judgments by the other experts, usually leading to convergence among the judgments. Research on group judgments has found that "group think" often prevails in groups with strong interactions, particularly in cohesive groups meeting face-to-face. This results in a misleading

convergence of opinion (McCauley, 1989) and exacerbates overconfidence (Gustafsen *et al.*, 1973). It is cased by a tendency to avoid conflict and the dominance of one or two participants to dominate the group, even though they do not have greater expertise. It is, therefore, usually better to obtain opinions from each expert individually rather than as a group. However, it can be useful to ask the experts to discuss relevant evidence as a group before they make their assessments. In this way, experts become aware of all the potentially relevant evidence, and may learn key strengths or shortcomings of evidence that they did not know.

There has been extensive theoretical research on techniques to combine probability distributions from multiple experts (Cooke, 1991). Much of it concerns ways to weight opinions and to model the probabilistic dependence among the experts. As a practical matter, it is hard to assess such dependence. Experts will usually have large if incomplete overlaps in their awareness of relevant research and evidence, but differing evaluations of the relevance and credibility of the evidence, even after sharing their views on the evidence. For these reasons, the sophisticated mathematical combination techniques are often hard to apply in practice. There has been some empirical evaluation comparing different combination methods, which suggests that simple weighted combination of distributions is usually as good as the more sophisticated methods (Cooke 1991).

Some studies weight the experts according to their degree of expertise. Asking experts to rate each others' expertise can be contentious, especially when there are strong differences of opinion. Instead, it is often best to ask experts to rate their own expertise—separately on each quantity since their expertise may vary.

If there is significant overlap among expert's distributions, the selection of weighting and combination method makes little difference to the results. But if some opinions have little overlap with each other—for example, when there are strongly differing schools of thought on the topic—it is often best not to combine the opinions at all. Instead, the analysis can be conducted separately for each school of thought.

To date, elicitation of individual experts has been the most widely used method of using expert judgment to characterize uncertainty about climate-related issues.

In that way, the effects of the differences of opinion are clear in the results, instead of papering them over.

Budnitz *et al.* (1995, 1998) have recently developed a much more elaborate group method in the context of probabilistic seismic hazard analysis. Meeting for an extended period, a group of experts work collectively, not as proponents of specific viewpoints but rather as:

> ...informed *evaluators* of a range of viewpoints. (These individual viewpoints or models may be defended by proponents experts invited to present their views and "debate" the panel). Separately the experts on the panel also play the role of *integrators*, providing advice...on the appropriate representation of the composite position of the community as a whole.

A technical facilitator/integrator (TFI):

> ...conducts both individual elicitations and group interactions, and with the help of the experts themselves the TFI integrates data, models and interpretations to arrive at the final product: a full probabilistic characterization of the seismic hazard at a site, including the uncertainty. Together with the experts acting as evaluators, the TFI "owns" the study and defends it as appropriate (Budnitz *et al.*, 1998).

Needless to say the process is very time consuming and expensive, requiring weeks or more of the experts' time.

Protocols for Individual Expert Elicitation

Developing a protocol for an effective expert elicitation in a substantively complex domain, such as climate science or climate impacts, typically requires many months of development, testing, and refinement[3]. Typically the designers of such protocols start with many more questions they would like to pose than experts are likely to have patience or the ability to answer. Iteration is required to reduce the list of questions to those most essential and to formulate

Developing a protocol for an effective expert elicitation in a substantively complex domain, such as climate science or climate impacts, typically requires many months of development, testing, and refinement.

questions of a form that is unambiguous and compatible with the way in which experts frame and think about the issues at hand. To achieve this latter, sometimes it is necessary to provide a number of different response modes. In this case, designers need to think about how they will process results to allow appropriate comparisons of different expert responses. To support this objective, it is often desirable to include some redundancy in the protocol enabling tests of the internal consistency of the experts' judgments.

A number of basic protocol designs have been outlined in the literature (see Chapter 7 in Morgan and Henrion [1990] and associated references). Typically, they begin with some explanation of why the study is being conducted and how the results will be used. In most cases, experts are told that their names will be made public but that their identity will not be linked to any specific answer. This is done to minimize the possible impact of peer pressure, especially in connection with requests to estimate extreme values. Next, some explanation is typically provided of the problems posed by cognitive heuristics and overconfidence. Some interviewers in the decision analysis community ask experts to respond to various "encyclopedia questions" or perform other exercises to demonstrate the ubiquitous nature of overconfidence in the hopes that this "training" will help to reduce overconfidence in the answers received. Unfortunately, the literature suggests that such efforts have little, if any, effect[4]. However, asking specific "disconfirming" questions, or "stretching" questions such as "Can you explain how the true value could turn out to be much larger (smaller) than your extreme value?" (see below) can be quite effective in reducing overconfidence.

In elicitations they have done on rather well-defined topics, Cooke (1991) and his colleagues[5] have placed considerable emphasis on checking expert calibration and performance by presenting them with related questions for which values are well known, and then giving greater weight to experts who perform well on those questions.

---

[3]   Roger Cooke (1991) and his colleagues have developed a number of elicitation programs in much shorter periods of time, working primarily in problem domains in which the problem is well specified and the specific quantities of interest are well defined.

[4]   See, for example, the discussion on pp. 120-122 of Morgan and Henrion (1990).
[5]   Additional information about some of this work can be found at <http://www.rff.org/Events/Pages/Expert-Judgment-Workshop-Documents.aspx>. See also Kurowicka and Cooke (2006).

Others in the decision science community are not persuaded that such weighting strategies are advisable.

While eliciting a cumulative density function (CDF) of a probability distribution to characterize the uncertainty about the value of a coefficient of interest is the canonical question form in expert elicitations, many of the elicitation protocols used in climate science have involved a wide range of other response modes (Morgan and Keith, 1995; Morgan *et al.*, 2001, 2006; Zickfeld *et al.*, 2007). In eliciting a CDF, it is essential to first clearly resolve with the expert exactly what quantity is being considered so as to remove ambiguity that might be interpreted differently by different experts. Looking back across a number of past elicitations, it appears that the uncertainty in question formulation and interpretation can sometimes be as large or larger than uncertainty arising from the specific formulation used to elicit CDFs. However, this is an uncertainty that can be largely eliminated with careful pilot testing, refinement, and administration of the interview protocol.

Once a clear understanding about the definition of the quantity has been reached, the usual practice is to begin by asking the expert to estimate upper and lower bounds. This is done in an effort to minimize the impact of anchoring and adjustment and associated overconfidence. After receiving a response, the interviewer typically then chooses a slightly more extreme value (or, if it exists, cites contradictory evidence from the literature) and asks if the expert can provide an explanation of how that more extreme value could occur. If an explanation is forthcoming, the expert is then asked to consider extending the bound. Only after the outer range of the possible values of the quantity of interest has been established does the interviewer go on to pose questions to fill in the balance of the distribution, using standard methods from the literature (Morgan and Henrion, 1990).

Experts often have great difficulty in thinking about extreme values. Sometimes they are more comfortable if given an associated probability (*e.g.*, a 1:100 upper bound rather than an absolute upper bound). Sometimes they give very different (much wider) ranges if explicitly asked to include "surprises", even though the task at hand has been clearly defined as identifying the range of all possible values. Therefore, where appropriate, the investigator should remind experts that "surprises" are to be incorporated in the estimates of uncertainty.

Hammitt and Shlyakhter (1999) have noted that overconfidence can give rise to an underestimate of the value of information in decision analytic applications. They note that because "the expected value of information depends on the prior distribution used to represent current uncertainty", and observe that "if the prior distribution is too narrow, in many risk-analytic cases, the calculated expected value of information will be biased downward". They have suggested a number of procedures to guard against this problem.

Most substantively detailed climate expert elicitations conducted to date have involved extended face-to-face interviews, typically in the expert's own office so that they can access reference material (and in a few cases even ask colleagues to run analyses, *etc.*). This has several clear advantages over mail or web-based methods. The interviewers can:
- have confidence that the expert is giving his or her full attention and careful consideration to the questions being posed and to performing other tasks;
- more readily identify and resolve confusion over the meaning of questions, or inconsistencies in an expert's responses;
- more easily offer conflicting evidence from the literature to make sure that the expert has considered the full range of possible views;
- build the greater rapport typically needed to pose more challenging questions and other tasks (such as ranking research priorities).

While developing probabilistic estimates of the value of key variables (*i.e.*, empirical quantities) can be extremely useful, it is often even more important to develop an understanding of how experts view uncertainty about functional relationships among variables. To date, this has received rather less attention in most elicitation studies; however, several have attempted to pose questions that address such uncertainties.

Experts often have great difficulty in thinking about extreme values. Therefore, where appropriate, the investigator should remind experts that "surprises" are to be incorporated in the estimates of uncertainty.

Propagation and Analysis of Uncertainty

Lead Author: M. Granger Morgan, Carnegie Mellon Univ.

Contributing Authors: Hadi Dowlatabadi, Univ. of British Columbia; Max Henrion, Lumina Decision Systems; David Keith, Univ. of Calgary; Robert Lempert, The RAND Corporation; Sandra McBride, Duke Univ.; Mitchell Small, Carnegie Mellon Univ.; Thomas Wilbanks, Oak Ridge National Laboratory

Probabilistic descriptions of what is known about some key quantities can have value in their own right as an input to research planning and in a variety of assessment activities. Often, however, analysts want to incorporate such probabilistic descriptions in subsequent modeling and other analyses. A number of closed-form analytical methods exist to perform uncertainty analysis (Morgan and Henrion, 1990). However, as computing power and speed have continued to grow, most of the standard methods for the propagation of uncertainty through models, and the analysis of its implications, have come to depend on stochastic simulation.

Such methods are now widely used in environmental, energy, and policy research, either employing standard analysis environments such as @risk® <http://www.atrisk.com>, Crystal Ball® <http://www.crystalball.com>, and Analytica® <http://www.lumina.com>, or writing special purpose software to perform such analyses.

While modern computer methods allow investigators to represent all model inputs as uncertain, and propagate them through all but the most computationally intensive models[1] using stochastic simulation, it is often useful to explore how much uncertainty each input variable contributes to the overall uncertainty in the output of the model. A number of methods are now available to support such an assessment, many of which have recently been reviewed and critiqued by Borgonovo (2006).

Many studies have used Nordhaus' simple DICE and RICE models (Nordhaus and Boyer, 2000) to examine optimal emissions abatement policies under uncertainty. In a more recent work, Keller *et al.* (2005) has used a modified version of the RICE model to examine the implications of uncertainty about potential abrupt collapse of the North Atlantic Meridian Overturning Circulation (Gulf Stream).

---

[1] These methods are routinely used on a wide variety of engineering-economic, environmental, and policy models. With present computational resources brute force stochastic simulation is not feasible on large atmospheric General Circulation Models (GCMs), although parametric methods can be used such as those employed by climateprediction.net (See: <http://climateprediction.net>).

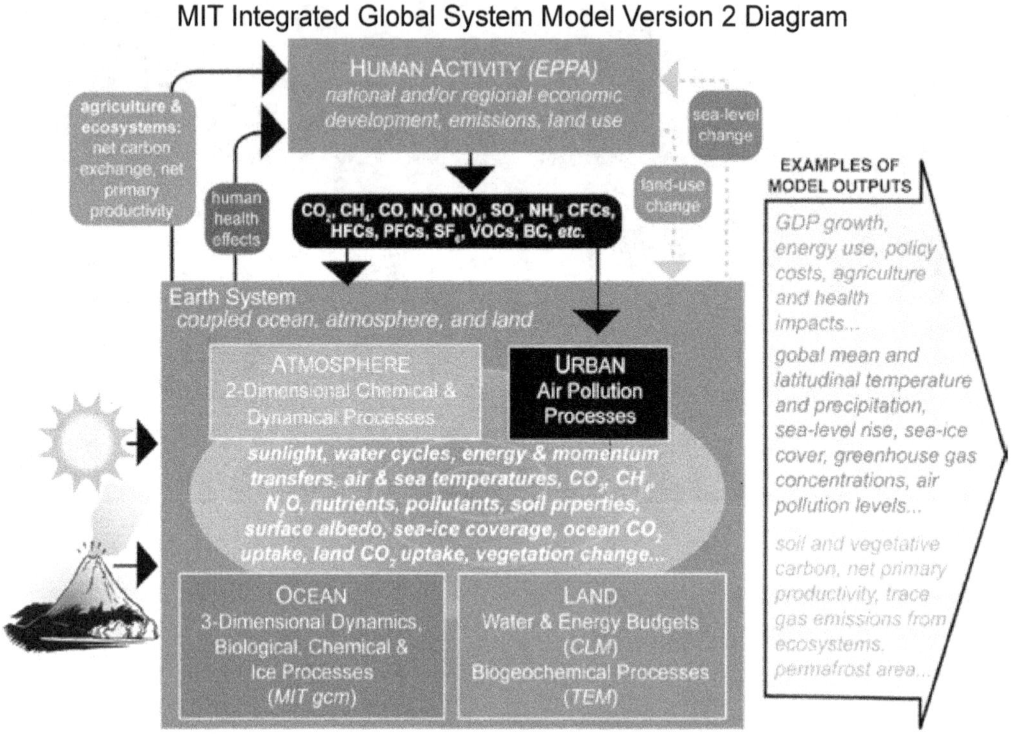

MIT Integrated Global System Model Version 2 Diagram

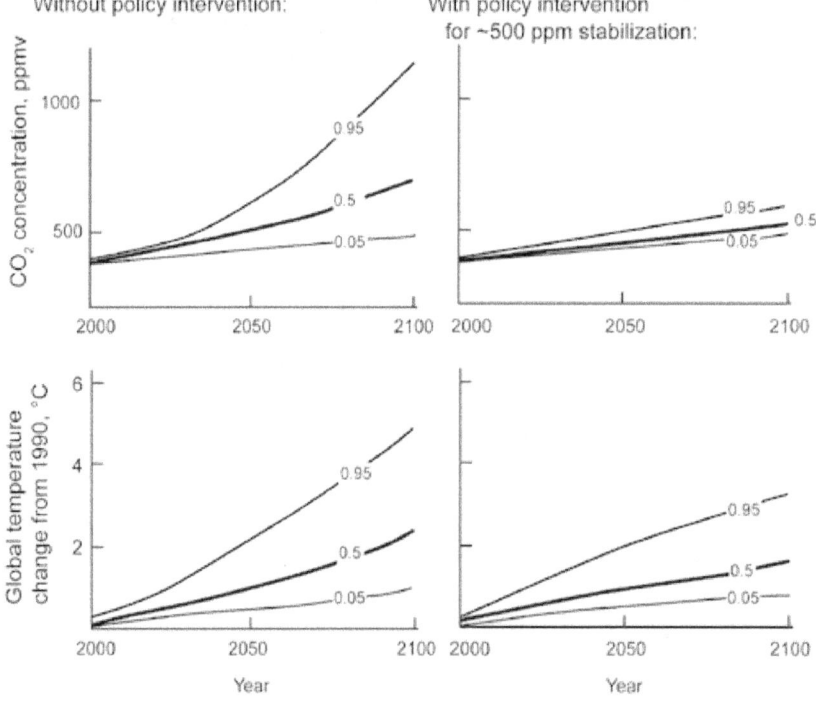

Figure 6.1 Simplified block diagram of the MIT Integrated Global System Model Version 2. Source: MIT Global Change Joint Program. Reprinted with permission.

MIT Integrated Global System Model Version 2 Simulation Results

Figure 6.2 Results of simulation conducted by Webster *et al.* (2003) that use an earlier version of the MIT IGSM model with probability distributions on model inputs that are constrained by past performance of the climate system. Results on the left are the authors' projection for no policy intervention and on the right for a specific policy intervention that roughly achieves stabilization at 500 parts per million by volume (ppmv). Heavy curves show median results from the simulations. Light curves show 0.05 and 0.95 confidence intervals. [Redrawn from Webster *et al.* (2003).]

Combining Expert Judgment and Model Simulations to Bound Sensitivity

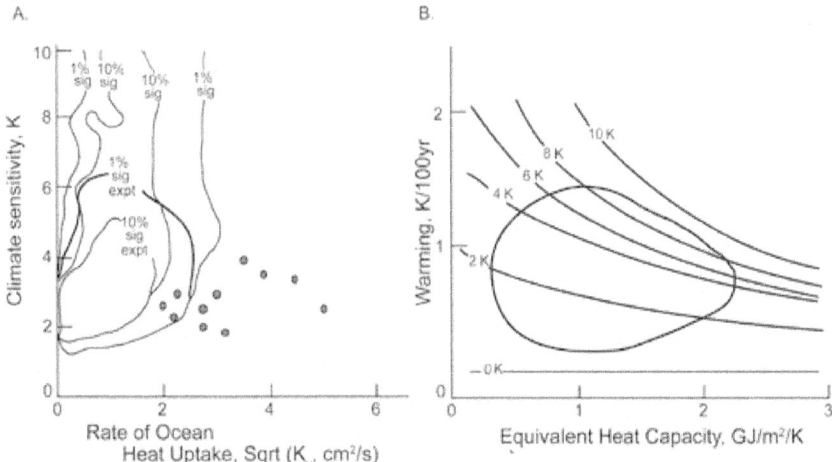

Figure 6.3 Two examples of recent efforts to bound sensitivity and heat uptake or heat capacity by combining expert judgment and model simulations. (a) (redrawn from Forest *et al.*, 2006) shows the marginal posterior probability density function obtained when using uniform probability distributions across all relevant forcings and matching outputs from the ocean and atmospheric portion of the MIT Integrated Global System Model. Light contours bound the 10 percent and 1 percent significance regions. Similarly, the two dark contours are for an expert probability density function on climate sensitivity. Dots show outputs from a range of leading general circulation models, all of which lie to the right of the high-probability region, suggesting that if Forest *et al.* (2006) are correct, these models may be mixing heat into the deep ocean too efficiently. (b) (redrawn from Frame *et al.*, 2005) shows the relationship between climate sensitivity, shown as light contours, effective ocean heat capacity, and twentieth century warming for the case of uniform sampling of climate sensitivity (not shown are similar results for uniform sampling across feedback strength). The dark contour shows the region consistent with observations at the five percent level. Note: We have roughly extrapolated the climate sensitivity contours from colored points in the original diagram that report each of many of hundreds of individual model runs. In this diagram, they are only qualitatively correct. Note that neither of these analyses account for the issue of uncertainty about model structural form.

Other groups, such as the Integrated Climate Assessment Model (ICAM) effort (Dowlatabadi and Morgan, 1993; Morgan and Dowlatabadi, 1996; Dowlatabadi, 2000) and the Massachusetts Institute of Technology (MIT) Joint Program[2], have propagated uncertainty through more complex integrated assessment models.

A description of the MIT Integrated Global System Model (IGSM) can be found in Sokolov *et al.* (2005) and on the web at <http://web.mit.edu/globalchange/www/if.html>. As shown in Figure 6.1, anthropogenic and natural emissions models are used to provide forcings for a coupled two-dimensional land- and ocean-resolving model of the atmosphere that is coupled to a three-dimensional ocean general circulation model. Outputs of that model are used as inputs to a terrestrial ecosystems model that predicts land vegetation changes, land carbon dioxide

(CO_2) fluxes, and soil composition. These in turn feed back to the coupled chemistry/climate and natural emissions models.

Webster *et al.* (2003) used an earlier version of the MIT model to perform a stochastic simulation that explores the uncertainty associated with a specific policy intervention that roughly achieves stabilization at 500 parts per million by volume (ppmv). Results are shown in Figure 6.2.

Using this and similar models, investigators associated with the MIT Joint Center have conducted a variety of uncertainty analyses. For example, Forest *et al.* (2002, 2006) have used an optimal fingerprinting method to bound the range of values of climate sensitivity and the rate of ocean heat uptake that are consistent with their model when matched with the observed climate record of the twentieth century. An example of a recent result is shown in Figure 6.3a.

---

[2] For a list of publications from the MIT Joint Program, see <http://web.mit.edu/globalchange/www/reports.html>.

Using a simple global energy balance model and diffusive ocean, Frame *et al.* (2005) have conducted studies to constrain possible values of climate sensitivity given plausible values of effective ocean heat capacity and observed twentieth century warming. An example result is shown in Figure 6.3b. The result shown is for uniform weighting across climate sensitivity. Uniform weighting across feedbacks yields somewhat different results. The authors note that their results "fail to obtain a useful upper bound on climate sensitivity unless it is assumed *a priori*".

Frame *et al.* (2005) conclude that:

...if the focus is on equilibrium warming, then we cannot rule out high sensitivity, high heat uptake cases that are consistent with, but non-linearly related to, 20th century observations. On the other hand, sampling parameters to simulate a uniform distribution of transient climate response...gives an approximately uniform distribution in much more immediately policy-relevant variables...under all SRES emission scenarios. After weighting for observations...this approach implies a 5-95% range of uncertainty in S [the climate sensitivity] of 1.2-5.2°C, with a median of 2.3°C, suggesting traditional heuristic ranges of uncertainty in S (IPCC, 2001) may have greater relevance to medium-term policy issues than recent more formal estimates based on explicit uniform prior distributions in either S or [feedback strength] λ.

Murphy *et al.* (2004) have completed extensive parametric analysis with the HadAM3 atmospheric model coupled to a mixed layer ocean that they report "allows integration to equilibrium in a few decades". They selected a subset of 29 of the roughly 100 parameters in this model, which were judged "by modelling experts as controlling key physical characteristics of sub-grid scale atmospheric and surface processes" and " perturbed these one at a time relative to the standard version of the GCM... creating a perturbed physics ensemble (PPE) of 53 model versions each used to simulate present-day and doubled CO_2 climates".

Placing uniform probability distributions on all these, they conclude that the implied climate sensitivity has a "median value of 2.9°C with a spread (corresponding to a 5 to 95% probability range) of 1.9 to 5.3°C". By using some analysis and expert judgment to shape the prior distributions, they also produce a "likelihood-weighted" distribution that they report "results in a narrowing of the 5 to 95% probability range to 2.4 to 5.4°C, while the median value increases to 3.5°C" (Murphy *et al.*, 2004). They report:

Our probability function is constrained by objective estimates of the relative reliability of different model versions, the choice of model parameters that are varied and their uncertainty ranges, specified on the basis of expert advice. Our ensemble produces a range of regional changes much wider than indicated by traditional methods based on scaling the response patterns of an individual simulation.

One of the most exciting recent developments in exploring the role of uncertainty in climate modeling has been the use of a large network of personal computers, which run a version of the HadSM3 model as a background program when machine owners are not making other uses of their machines. This effort has been spearheaded by Myles Allen and colleagues at Oxford (Allen, 1999). Details can be found at <http://www.climateprediction.net/index.php>. As of mid-spring 2006, this network involved over 47,000 participating machines that had completed over 150,000 runs of a version of the HadSM3 model, for a total of 11.4 million model years of simulations.

Initial results from this work were reported by Stainforth *et al.* (2005), who summarize their findings from a study of 2,578 simulations of the model as follows:

We find model versions as realistic as other state-of-the-art climate models but with climate sensitivities ranging from less than 2K to more than 11K. Models with such extreme sensitivities are critical for the study of the full range of possible responses of the climate system to rising greenhouse gas levels, and for assessing the risks associated with

a specific target for stabilizing these
levels...

The range of sensitivity across different
versions of the same model is more than
twice that found in the GCMs used in
the IPCC Third Assessment Report...
The possibility of such high sensitivities has been reported by studies using
observations to constrain this quantity,
but this is the first time that GCMs have
generated such behavior.

The frequency distribution in climate sensitivity they report across all model versions is
shown in Figure 6.4.

Annan and Hargreaves (2006) have used Bayes'
Theorem and a set of likelihood functions that
they constructed for twentieth century warming,
volcanic cooling, and cooling during the last
glacial maximum, to "...conclude that climate
sensitivity is very unlikely (< 5% probability)
to exceed 4.5°C" and to argue that they "...can
not assign a significant probability to climate
sensitivity exceeding 6°C without making what
appear to be wholly unrealistic exaggerations
about the uncertainties involved".

While the common practice in many problem
domains is to build predictive models, or perform various forms of policy optimization, it is
important to ask whether meaningful prediction
is possible. Roe and Baker (2007) have argued
that, in the context of climate sensitivity, better
understanding of the operation of individual
physical processes may not dramatically improve one's ability to estimate the value of
climate sensitivity:

> We show that the shape of these prob-
> ability distributions is an inevitable and
> general consequence of the nature of the
> climate system, and we derive a simple
> analytic form for the shape that fits re-
> cent published distributions very well.
> We show that the breadth of the distribu-
> tion and, in particular, the probability of
> large temperature increases are relatively
> insensitive to decreases in uncertainties
> associated with the underlying climate
> processes.

Histogram: Climate Sensitivities

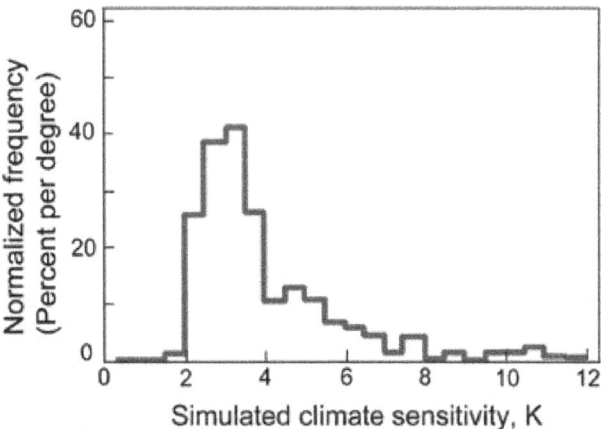

Figure 6.4 Histogram (redrawn) of climate sensitivities found by Stainforth
et al. (2005) in their simulation of 2,578 versions of the HadSM3 general
circulation model.

U.S. Primary Energy Consumption Forecasts

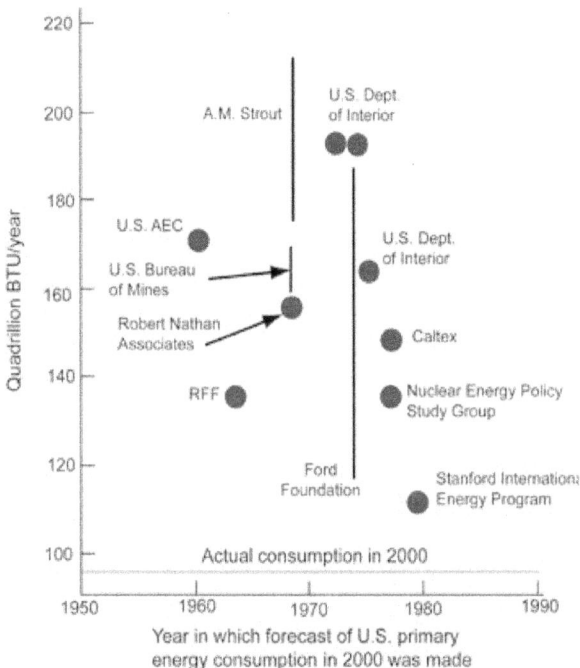

Figure 6.5 Summary of forecasts of United States primary energy
consumption compiled by Smil (2003) as a function of the date on which
they were made. (Figure redrawn from Smil, 2003.)

In the context of predicting the future evolution
of the energy system, which is responsible for a
large fraction of anthropogenic greenhouse gas
emissions, Smil (2003) and Craig *et al.* (2002)
have very clearly shown that accurate prediction
for more than a few years in the future is virtually impossible. Figure 6.5, redrawn from Smil

(2003), shows the sorry history of past forecasts for U.S. energy consumption. His summary of forecasts of global energy consumption shows similarly poor performance.

In addition to uncertainties about the long-term evolution of the energy system and hence future emissions, uncertainties about the likely response of the climate system, and about the possible impacts of climate change, are so great that a full characterization of coefficient and model uncertainty in a simulation model can lead to probabilistic results that are so broad that they are effectively useless (Casman *et al.*, 1999). Similarly, if one does parametric analysis across different model formulations, one can obtain an enormous range of answers depending on the model form and other inputs that are chosen. This suggests that there are decided limits to the use of "predictive models" and "optimization" in many climate assessment and policy settings.

The difficulties, or sometimes even impossibility, of performing meaningful predictive analysis under conditions of what has been called "deep" or "irreducible" uncertainty have led some investigators to pursue a different approach based on two key ideas: describing uncertainty about the system relevant to a decision with multiple representations, as opposed to a single best-estimate joint probability distribution, and using a robustness, as opposed to an optimality, as the criteria for evaluating alternative policy options. We turn to a more detailed discussion of these approaches in the latter parts of Part 7.

Making Decisions in the Face of Uncertainty

Lead Author: M. Granger Morgan, Carnegie Mellon Univ.

Contributing Authors: Hadi Dowlatabadi, Univ. of British
Columbia; Max Henrion, Lumina Decision Systems; David Keith,
Univ. of Calgary; Robert Lempert, The RAND Corporation; Sandra
McBride, Duke Univ.; Mitchell Small, Carnegie Mellon Univ.; Thomas
Wilbanks, Oak Ridge National Laboratory

As we noted in Part 1, there are a number of things that are different about the climate problem (Morgan *et al.*, 1999), but high levels of uncertainty is not one of them. In our private lives, we decide where to go to college, what job to take, whom to marry, what home to buy, when and whether to have children, and countless other important choices, all in the face of large, and often irreducible, uncertainty. The same is true of decisions made by companies and by governments— sometimes because decisions must be made, sometimes because scientific uncertainties are not the determining factor (*e.g.*, Wilbanks and Lee, 1985), and sometimes because strategies can be identified that incorporate uncertainties and associated risks into the decision process (NRC, 1986).

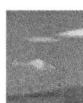

Classical decision analysis provides an analytical strategy for choosing among options when possible outcomes, their probability of occurrence, and the value each holds for the decision maker, can be specified. Decision analysis identifies an "optimal" choice among actions. It is rigorously derived from a set of normatively appealing axioms (Raiffa and Schlaifer, 1968; Howard and Matheson, 1977; Keeney, 1982). In applying decision analysis, one develops and refines a model that relates the decision makers' choices to important outcomes. One must also determine the decision maker's utility function(s)[1] in order to determine which outcomes are most desirable. One then propagates the uncertainty in various input parameters through the model (appropriately accounting for possible correlation structures among uncertain variables) to generate the expected utility of the various choice options. The best option is typically assumed to be the one with the largest expected utility, although other decision rules are sometimes employed.

---

[1] Many economists and analysts appear to assume that fully articulated utility functions exist in peoples' heads for all key outcomes, and that determining them is a matter of measurement. Many psychologists, and some decision analysts, suggest that this is often not the case and that for many issues people need help in thinking through and constructing their values (von Winterfeldt and Edwards, 1986; Fischhoff, 1991, 2005; Keeney, 1992).

In most cases, more research reduces uncertainty. In complex problems, such as some of the details of climate science, many years, or even many decades may go by, during which one's understanding of the problem grows richer, but the amount of uncertainty, remains unchanged, or even grows larger because research reveals processes or complications that had not previously been understood or anticipated.

When the uncertainty is well characterized and the model structure well known, this type of analysis can suggest the statistically optimal strategy to decision makers. Because there are excellent texts that outline these methods in detail (*e.g.*, Hammond *et al.*, 1999), we do not elaborate the ideas further here.

In complex and highly uncertain contexts, such as those involved in many climate-related decisions, the conditions needed for the application of conventional decision analysis sometimes do not arise (Morgan *et al.*, 1999). Where uncertainty is large, efforts can be made to reduce the uncertainties—in effect, reducing the width of probability distributions through research to understand underlying processes better. Alternatively, efforts can be made to improve understanding of the uncertainties themselves so that they can be more confidently incorporated in decision-making strategies.

In most cases, more research reduces uncertainty. Classic decision analysis implicitly assumes that research always reduces uncertainty. While eventually it usually does, in complex problems, such as some of the details of climate science, many years, or even many decades may go by, during which one's understanding of the problem grows richer, but the amount of uncertainty, as measured by our ability to make specific predictions, remains unchanged, or even grows larger because research reveals processes or complications that had not previously been understood or anticipated. That climate experts understand this is clearly demonstrated in the results from Morgan and Keith (1995) shown in Table 7.1. Unfortunately, many others do not recognize this fact, or choose to ignore it in policy discussions. This is not to argue that research in understanding climate science, climate impacts, and the likely effectiveness of various climate management policies and technologies is not valuable. Clearly it is. But when it does not immediately reduce uncertainty we should remember that there is also great value in learning that we knew less than we thought we did. In some cases, all the research in the world may not eliminate

key uncertainties on the timescales of decisions we must make[2].

In the expert elicitations of climate scientists conducted by Morgan and Keith (1995), experts were asked to design a 15-year long research program funded at a billion dollars per year that was designed to reduce the uncertainty in our knowledge of climate sensitivity and related issues. Having done this, the experts were asked how much they thought their uncertainty might have changed if they were asked the same ques-

Table 7.1 In the expert elicitations of climate scientists conducted by Morgan and Keith (1995), experts were asked to design a 15-year long research program funded at a billion dollars per year that was designed to reduce the uncertainty in our knowledge of climate sensitivity and related issues. Having done this, the experts were asked how much they thought their uncertainty might have changed if they were asked the same question in 15 years. The results below show that like all good scientists the experts understand that research does not always reduce uncertainty. Note: Expert 3 used a different response mode for this question. He gave a 30 percent increase by a factor of less than or equal to 2.5.

Expert Number	Chance that the experts believe that their uncertainty about the value of climate sensitivity would *grow* by greater than 25 percent after a 15-year 10^9 per year research program
1	10
2	18
3	30 (Note)
4	22
5	30
6	14
7	20
8	25
9	12
10	20
11	40
12	16
13	12
14	18
15	14
16	8

[2] In general we believe that if policy makers are made aware of the nature of uncertainty and the potential for its reduction, they will be in a position to make better decisions.

Decision Strategies

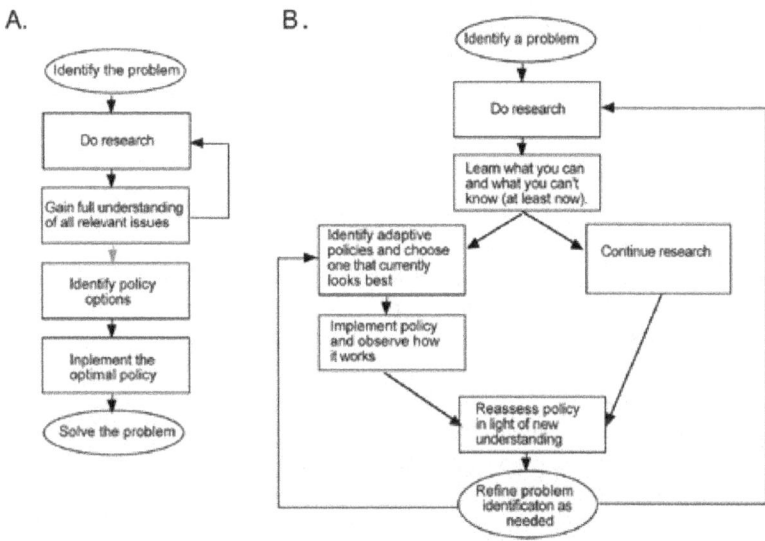

Figure 7.1 In the face of high levels of uncertainty, which may not be readily resolved through research, decision makers are best advised to not adopt a decision strategy in which (a) nothing is done until research resolves all key uncertainties, but rather (b) to adopt an iterative and adaptive strategy.

tion in 15 years. The results below show that like all good scientists the experts understand that research does not always reduce uncertainty. Note: Expert 3 used a different response mode for this question. He gave a 30 percent increase by a factor of less than or equal to 2.5.

This raises the question of what considerations should drive research. Not all knowledge is likely to be equally important in the climate-related decisions that individuals, organizations, and nations will face over the coming decades. Thus, while it is often hard to do (Morgan *et al.*, 2006), when possible, impact assessors, policy analysts, and research planners should consider working backward from the decisions they face to design research programs which are most likely to yield useful insights and understanding.

There are two related decision-making/management strategies that may be especially appealing in the face of high uncertainty. These are:
Resilient Strategies: In this case, the idea is to try to identify the range of future circumstances that one might face, and then seek to identify approaches that will work reasonably well across that range.

Adaptive Strategies: In this case, the idea is to choose strategies that can be modified to achieve better performance as one learns more about the issues at hand and how the future is unfolding.

Both of these approaches stand in rather stark contrast to the idea of developing optimal strategies that has characterized some of the work in the integrated assessment community, in which it is assumed that a single model accurately reflects the nature of the world, and the task is to choose an optimal strategy in that well-specified world.

The ideas of resilience and adaptation have been strongly informed by the literature in ecology. Particularly good discussions can be found in Clark (1980) and Lee (1993). A key feature of adaptive strategies is that decision makers learn whatever they can about the problem they face and then make choices based on their best assessment and that of people whose advice they value. They seek strategies that will let them, or those who come after them, modify choices in accordance with insights gained from more experience and research. That is, rather than adopt a decision strategy of the sort shown in Figure 7.1a in which nothing is done until research resolves all key uncertainties; they adopt

When possible, impact assessors, policy analysts, and research planners should consider working backward from the decisions they face to design research programs which are most likely to yield useful insights and understanding.

an iterative and adaptive strategy that looks more like that shown in Figure 7.1b. Adaptive strategies work best in situations in which there are not large non-linearities and in which the decision time scales are well matched to the changes being observed in the world.

A familiar example of a robust strategy is portfolio theory as applied in financial investment, which suggests that greater uncertainty (or a lesser capacity to absorb risks) calls for greater portfolio diversification. Another example arose during the first regional workshop conducted by the National Assessment Synthesis Team in Fort Collins, Colorado, in preparation for developing the U.S. National Climate Change Assessment (NAST, 2001). Farmers and ranchers participating in the discussion suggested that, if possible climate change introduces new uncertainties into future climate forecasts, it might be prudent for them to reverse a trend toward highly-specialized precision farming and ranching, moving back toward a greater variety of crops and range grasses.

Deep Uncertainty

Decision makers face deep uncertainty when those involved in a decision do not know or cannot agree upon the system model that relates actions to consequences or the prior probability distributions on the input parameters to any system model[3]. Under such conditions, multiple representations can provide a useful description of the uncertainty.

Most simply, one can represent deep uncertainty about the values of empirical quantities and about model function form by considering multiple cases. This is the approach taken by traditional scenario analyses. Such traditional scenarios present a number of challenges, as documented by Parson *et al.* (2007). Others have adopted multi-scenario simulation approaches (IPCC, 2001) where a simulation model is run many times to create a large number of fundamentally different futures and used directly to

make policy arguments based on comparisons of these alternative cases.

In the view of the authors of this Product, considering a set of different, plausible joint probability distributions over the input parameters to one or more models provides the most useful means to describe deep uncertainty. As described below, this approach is often implemented by comparing the ranking or desirability of alternative policy decisions as a function of alternative probability weightings over different states of the world. This is similar to conventional sensitivity analysis where one might vary parameter values or the distribution over the parameters to examine the effects on the conclusions of an analysis. However, the key difference is one of degree. Under deep uncertainty, the set of plausible distributions contains members that in fact would imply very different conclusions for the analysis. In addition to providing a useful description of deep uncertainty, multiple representations can also play an important role in the acceptance of the analysis when stakeholders to a decision have differing interests and hold differing, non-falsifiable, perceptions. In such cases, an analysis may prove more acceptable to all sides in a debate if it encompasses all the varying perspectives rather than adopting one view as privileged or superior (Rosenhead and Mingers, 2001).

There exists no single definition of robustness. Some authors have defined robust strategy as one that performs well, compared to the alternatives, over a very wide range of alternative futures (Lempert *et al.* 2003). This definition represents a "satisficing" criterion (Simon, 1959), and is similar to domain criteria (Schneller and Sphicas, 1983) where decision makers seek to reduce the interval over which a strategy performs poorly. Another formulation defines a robust strategy as one that sacrifices a small amount of optimal performance in order to obtain less sensitivity to broken assumptions. This robustness definition underlies Ben-Haim's (2001) "Info-Gap" approach, the concept of robustness across competing models used in monetary policy applications (Levin and Williams, 2003), and to treatments of low-probability-high-consequence events (Lempert *et al.*, 2002). This definition draws on the observation that an optimum strategy

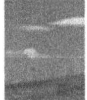

Decision makers face deep uncertainty when those involved in a decision do not know or cannot agree upon the system model that relates actions to consequences or the prior probability distributions on the input parameters to any system model.

---

[3] A number of different terms are used for what we call here 'deep uncertainty'. Knight (1921) distinguished risk from uncertainty, using the latter to denote factors poorly described by quantified probabilities. Ben-Haim (2001) refers to severe uncertainty and Vercelli (1994) to hard as opposed to the more traditional soft uncertainty. The literature on imprecise probabilities refers to probabilities that can lie within a range.

may often be brittle, that is, its performance may degrade rapidly under misspecification of the assumptions, and that decision makers may want to take steps to reduce that brittleness[4]. For instance, if one has a best-estimate joint probability distribution describing the future, one might choose a strategy with slightly less than optimal performance in order to improve the performance if the tails of the best-estimate distribution describing certain extreme cases turn out to be larger than expected[5]. Other authors have defined robustness as keeping options open. Rosenhead (2001) views planning under deep uncertainty as a series of sequential decisions. Each decision represents a commitment of resources that transforms some aspect of the decision maker's environment. A plan foreshadows a series of decisions that it is anticipated will be taken over time. A robust step is one that maximizes the number of desirable future end states still reachable, and, in some applications, the number of undesirable states not reachable, once the initial decision has been taken.

These definitions often suggest similar strategies as robust, but to our knowledge, there has been no thorough study that describes the conditions where these differing robustness criteria lead to similar or different rankings of alternative policy options. Overall, a robustness criterion often yields no single best answer but rather helps decision makers to use available scientific and socio-economic information to distinguish a set of reasonable choices from unreasonable choices and to understand the tradeoffs implied by choosing among the

reasonable options. Robustness can be usefully thought of as suggesting decision options that lie between an optimality and a minimax solution. In contrast to optimal strategies that, by definition, focus on the middle range of uncertainty most heavily weighted by the best estimate probability density function, robustness focuses more on, presumably unlikely but not impossible, extreme events and states of the world, without letting them completely dominate the decision.

One common means of achieving robustness is via an adaptive strategy, that is, one that can evolve over time in response to new information. Two early applications of robust decision making to greenhouse gas mitigation policies focused on making the case for such robust adaptive strategies. These studies also provide an example of a robust strategy as one that performs well over a wide range of futures. Morgan and Dowlatabadi (1996) used variants of their ICAM-2 model in an attempt to determine the probability that specific carbon tax policy would yield net positive benefits. Their sensitivity analysis over different model structures suggested a range that is so wide, 0.15 to 0.95, as to prove virtually useless for policy purposes. Similarly, Table 7.2 illustrates the wide range of effects due to alternative Integrated Climate Assessment Model (ICAM) structures one finds on the costs of carbon dioxide (CO_2) stabilization at 500 parts per million (ppm) (Dowlatabadi, 1998). To make sense of such deep uncertainty, Casman *et al.* (1999) considered adaptive decision strategies (implemented in the model as decision agents) that would take initial actions based on the current best forecasts, observe the results, revise their forecasts, and adjust their actions accordingly. This study highlights the importance of how we can build in robust strategies by building policies around different state variables. For example, the most common state variable in climate policy is annual emissions of greenhouse gases (GHGs). This variable suffers from high variability induced by: stochastic economic activity, energy market speculations, and inter-annual variability in climate. All of these factors can drive emissions up or down, outside the influence of the decision variable itself or how it influences the system (*i.e.*, a shadow price for GHGs). A policy that uses atmospheric concentration of CO_2 and its

> A robustness
> criterion helps
> decision makers use
> available scientific
> and socio-economic
> information
> to distinguish
> reasonable from
> unreasonable choices.

---

[4] Former United States Federal Reserve Chairman Alan Greenspan described an approach to robust strategies when he wrote "...For example, policy A might be judged as best advancing the policymakers' objectives, conditional on a particular model of the economy, but might also be seen as having relatively severe adverse consequences if the structure of the economy turns out to be other than the one assumed. On the other hand, policy B might be somewhat less effective under the assumed baseline model... but might be relatively benign in the event that the structure of the economy turns out to differ from the baseline. These considerations have inclined the Federal Reserve policymakers toward policies that limit the risk of deflation even though the baseline forecasts from most conventional models would not project such an event".

[5] Given a specific distribution, one can find a strategy that is optimal. But this is not the same as finding a strategy that performs well (satisfices) over a wide range of distributions and unknown system specifications.

Table 7.2 Illustration from Casman et al. (1999) of the wide range of results that can be obtained with the Integrated Climate Assessment Model (ICAM) depending upon different structural assumptions, in this case, about the structure of the energy module and assumptions about carbon emission control. In this illustration, produced with a 1997 version of the ICAM, all nations assume an equal burden of abatement by having a global carbon tax. Discounting is by a method proposed by Schelling (1995). Other versions of the ICAM yield qualitatively similar results.

Model Components		Model Variants								
		M1	M2	M3	M4	M5	M6	M7	M8	M9
Are new fossil oil & gas deposits discovered?		no	yes	no	no	yes	yes	no	yes	yes
Is technical progress that uses energy affected by fuel prices and carbon taxes?		no	no	yes	no	yes	yes	yes	yes	yes
Do the costs of abatement and non-fossil energy technologies fall as users gain experience?		no	no	no	yes	no	no	yes	yes	yes
Is there a policy to transfer carbon saving technologies to non Annex 1 countries?		no	no	no	no	no	yes	yes	no	yes
TPE BAU in 2100(EJ)	Mean	1975	2475	2250	2000	3425	2700	1450	3550	2850
TPE control in 2100 (EJ)	Mean	650	650	500	750	500	500	675	750	725
CO_2 BAU 2100 (10^9TC)	Mean	40	50	50	40	75	55	25	73	55
	Standard Deviation	28	18	36	29	29	23	22	27	21
Mitigation Cost (%Welfare)	Mean	0.23	0.44	0.14	0.12	0.48	0.33	0.05	0.23	0.17
	Standard Deviation	.045	0.23	0.23	0.22	0.28	0.12	0.07	0.12	0.11
Impact of delay (%Welfare)	Mean	-0.1	0.2	-0.6	0.0	-1	-0.5	-0.1	-0.6	-0.4
	Standard Deviation	1	0.3	1	0.7	1.2	0.9	0.5	0.8	0.6

Notes: TPE = Total Primary Energy
BAU = Business as Usual (no control and no intervention).
Sample size in ICAM simulation = 400

rate of change is much less volatile and much better at offering a robust signal for adjusting the decision variable through time. The study reports that atmospheric forcing, or GHG concentrations, is far more robust than alternative state variables such as emission rates or global average temperature over a wide range of model structures and parameter distributions. This finding has important implications for the types of scientific information that may prove most useful to decision makers.

Similarly, Lempert et al. (1996) used a simple integrated assessment model to examine the expectations about the future that would favor alternative emissions-reduction strategies. The study examined the expected net present value of alternative strategies as a function of the likelihood of large climate sensitivity, large climate impacts, and significant abatement-cost-reducing new technology. Using a policy region analysis (Watson and Buede, 1987), the study found that both a business-as-usual and a steep emissions-reduction strategy that do not adjust over time presented risky choices because they could prove far from optimal if the future turned out differently than expected. The study then compared an adaptive strategy that began with moderate initial emissions reductions and set specific thresholds for large future climate impacts and low future abatement costs. If the observed trends in impacts or costs trigger either threshold, then emissions reductions accelerate. As shown in Figure 7.2, this adaptive strategy performed better than the other two strategies over a very wide range of expectations about the future. It also proved to be close to optimal otherwise. For those expectations where one of the other two strategies performed best, the adaptive strategy performed nearly as well. The study thus concluded the adaptive decision strategy was robust compared to the two non-adaptive alternatives.

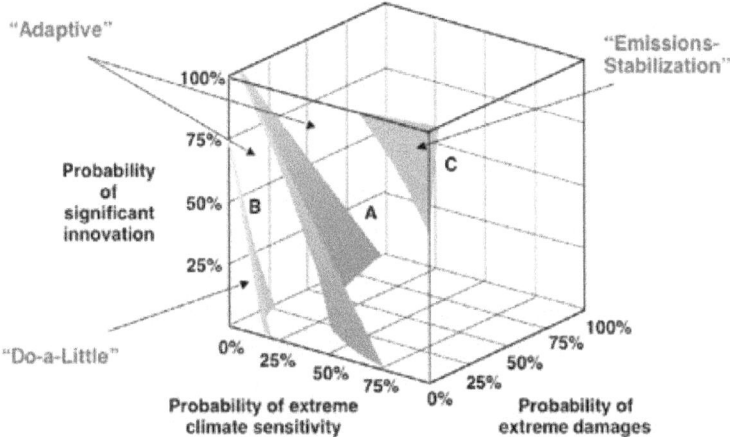

Figure 7.2 Surfaces separating the regions in probability space where the expected value of the "Do-A-Little" policy is preferred over the "Emissions-Stabilization" policy, the adaptive strategy is preferred over the "Do-A-Little" policy, and the adaptive strategy is preferred over the "Emissions-Stabilization" policy, as a function of the probability of extreme damages, significant innovation, and extreme climate sensitivity (Lempert *et al.*, 1996).

These robust decision-making approaches have been applied more recently using more sophisticated methods. For instance, Groves (2006) has examined robust strategies for California water policy in the face of climate and other uncertainties, and Dessai and Hulme (2007) have applied similar approaches to water resource management in the United Kingdom. Similarly, Hall (Hine and Hall, 2007) has used Haim's Info-Gap approach to examine robust designs for the Thames flood control system in the face of future scientific uncertainty about sea-level rise.

Surprise

Recent attention to the potential for abrupt climate change has raised the issue of "surprise" as one type of uncertainty that may be of interest to decision makers. An abrupt or discontinuous change represents a property of a physical or socio-economic system. For instance, similar to many such definitions in the literature, the United States National Academy of Sciences has defined an abrupt climate change as a change that occurs faster than the underlying driving forces (NRC, 2002). In contrast, surprise represents a property of the observer. An event becomes a surprise when it opens a significant gap between perceived reality and one's expectations (van Notten *et al.*, 2005; Glantz *et al.*, 1998; Hollings, 1986; Schneider *et al.*, 1998).

A number of psychological and organizational factors make it more likely that a discontinuity will cause surprise. For instance, individuals will tend to anchor their expectations of the future based on their memories of past patterns and observations of current trends and thus be surprised if those trends change. Scientists studying future climate change will often find a scarcity of data to support forecasts of systems in states far different than the ones they can observe today. Thus, using the taxonomy of Figure 1.1, the most well established scientific knowledge may not include discontinuities. For example, the sea-level rise estimates of the most recent Intergovernmental Panel on Climate Change (IPCC) Fourth Assessment Report (IPCC, 2007) do not include the more speculative estimates of the consequences of a collapse of the Greenland ice sheet because scientists' understanding of such a discontinuous change is less well-developed than for other processes of sea-level rise. Planners who rely only on the currently well-established estimates may come to be (or leave their successors) surprised. An analogy with earthquakes may be useful here[6]. Earthquakes are a well-known and unsurprising phenomenon, but a specific large quake at a specific time is still a big surprise for those hit by it since these cannot be forecast. One can build for earthquakes, but may choose not to do

---

[6] We thank Steven Sherwood of Yale University for this analogy and text.

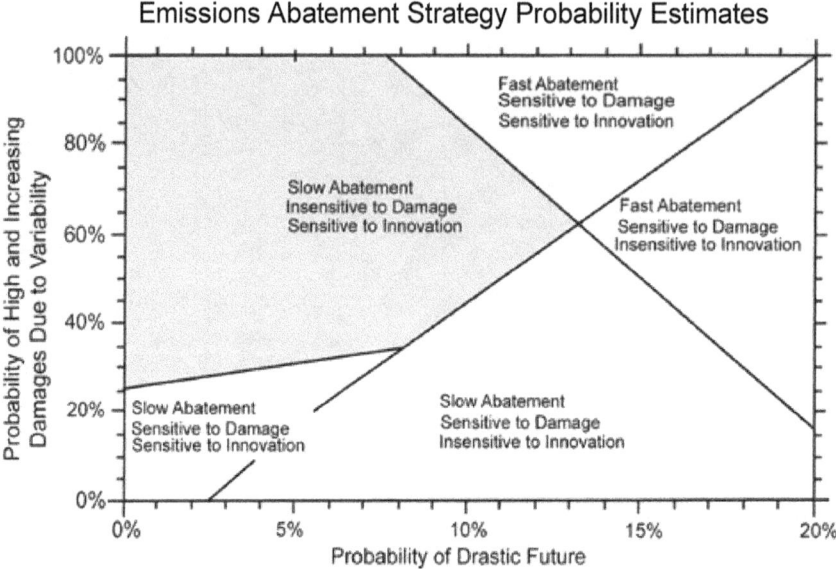

Figure 7.3 Estimates of the most robust emissions abatement strategy as a function of expectations about two key uncertainties—the probability of large future climate impacts and large future climate variability (Lempert and Schlesinger, 2006). Strategies are described by near-term abatement rate and the near-term indicators used to signal the need for any change in abatement rate. The shaded region characterizes range of uncertainty over which one strategy of interest is robust.

so in places not thought to be seismically active, although earthquakes even in such places are not unknown (*i.e.*, genuine surprises). It is very unlikely that we will ever be able to forecast in advance the moment when a particular ice sheet will collapse, until the unmistakable and irreversible signs of this are observed like the p-wave that arrives before the earthquake.

The concepts of robustness and resilience provide a useful framework for incorporating and communicating scientific information about potential surprise[7]. First, these concepts provide a potential response to surprise in addition to and potentially more successful than trying to predict them. A robust strategy is designed to perform reasonably well in the face of a wide range of contingencies and, thus, a well-designed strategy will be less vulnerable to a wide range of potential surprises whether predicted or not. Second, the robustness framework aims to provide a context that facilitates constructive consideration of otherwise unexpected events

---
[7] Robustness and resilience are related concepts. The former generally refers to strategies chosen by decision makers while the latter is a property of systems. However, the concepts overlap because decision makers can take actions that make a system more resilient.

(Lempert *et al.*, 2003). In general, there is no difficulty imagining a vast range of potential outcomes that might be regarded as surprising. It is in fact rare to experience a major surprise that had not been previously imagined by someone (*e.g.*, fall of the Soviet Union, Hurricane Katrina, Pearl Harbor, 9/11). The difficulty arises in a decision-making context if, in the absence of reliable predictions, there is no systematic way to prioritize, characterize, and incorporate the plethora of potential surprises that might be imagined. A robust decision framework can address this problem by focusing on the identification of those future states of the world in which a proposed robust strategy would fail, and then identify the probability threshold such a future would have to exceed in order to justify a decision maker taking near-term steps to prevent or reduce the impacts of such a future.

For example, Figure 7.3 shows the results of an analysis (Lempert *et al.*, 2000) that attempted to lay out the surprises to which a candidate emissions-reduction strategy might prove vulnerable. The underlying study considered the effects of uncertainty about natural climate variability on the design of robust, near-term emissions-mitigation strategies. This uncertainty about the level of natural variability makes it more difficult to determine the extent to which any observed climate trend is due to human-caused effects and, thus, makes it more difficult to set the signposts that would suggest emissions mitigation policies ought to be adjusted. The study first identified a strategy robust over the commonly discussed range of uncertainty about the potential impacts of climate change and the costs of emissions mitigation. It then examined a wider range of poorly characterized uncertainties in order to find those uncertainties to which the candidate robust strategy remains most vulnerable. The study finds two such uncertainties most important to the strategies' performance: the probability of unexpected large damages due to climate change and the probability of unexpectedly low damages due to changes in climate variability. Figure 7.3 traces the range of probabilities for these two uncertainties that would justify abandoning the proposed robust strategy described in the shaded region in favor of one of the other strategies shown on the figure. Rather than asking scientists or decision makers to quantify the probability of surprisingly large climate impacts, the analysis suggests that such a surprise would need to have a probability larger than roughly 10 to 15 percent in order to significantly influence the type of policy response the analysis would recommend. Initial findings suggest that this may provide a useful framework for facilitating the discovery, characterization, and communication of potential surprises.

Behavioral Decision Theory

The preceding discussion has focused on decision making by "rational actors". In the case of most important real-world decision problems, there may not be a single decision maker; decisions get worked out and implemented through organizations, in most cases formal analysis plays a subsidiary role to other factors, and in some cases, emotion and feelings (what psychologists term "affect") may play an important role.

These factors are extensively discussed in a set of literatures typically described as "behavioral decision theory" or risk-related decision making. In contrast to decision analysis that outlines how people should make decisions in the face of uncertainty if they subscribe to a number of axioms of rational decision making; these lit-

eratures are descriptive, describing how people actually make decisions when not supported by analytical procedures such as decision analysis. Good summaries can be found in Kahneman *et al.* (1982), Jaeger *et al.* (1998), and Hastie and Dawes (2001). Recently investigators have explored how rational and emotional parts of human psyche interact in decision making (Slovic, *et al.*, 2004; Peters *et al.*, 2006; Loewenstein *et al.*, 2001; Lerner *et al.*, 2003; Lerner and Tiedens, 2006). Far from diminishing the role of affect-based decision making, several of these authors argue that in many decision settings it can play an important role along with more analytical styles of thought.

There are also very large literatures on organizational behavior. One of the more important subsets of that literature for decision making under uncertainty concerns the processes by which organizational structure can play a central role in shaping the success of an organization in coping with uncertainty and strategies they can adopt to make themselves less susceptible to failure (see for example: LaPorte and Consolini, 1991; Vaughan, 1996; La Porte, 1996; Paté-Cornell *et al.*, 1997; Pool, 1997; Weick and Sutcliffe, 2001).

The "precautionary principle" is a decision strategy often proposed for use in the face of high uncertainty. There are many different notions of what this approach does and does not entail. In some forms, it incorporates ideas of resilience or adaptation. In some forms, it can also be shown to be entirely consistent with a decision analytic problem framing (DeKay *et al.*, 2002).

However, among some proponents, precaution has often taken the form of completely avoiding new activities or technologies that might hold the potential to cause adverse impacts, regardless of how remote their probability of occurrence. In this form, the precautionary principle has drawn vigorous criticism from a number of commentators. For example Sunstein (2005) argues:

> ...a wide variety of adverse effects may come from inaction, regulation and everything in between. [A better approach]...would attempt to consider all of these adverse effects, not simply

In the case of most important real-world decision problems, there may not be a single decision maker; decisions get worked out and implemented through organizations, in most cases formal analysis plays a subsidiary role to other factors, and in some cases, emotion and feelings may play an important role.

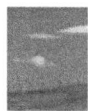

a subset. Such an approach would pursue distributional goals directly by, for example, requiring wealthy countries—the major contributors to the problem of global warming—to pay poor countries to reduce greenhouse gases or to prepare themselves for the relevant risks. When societies face risks of catastrophe, even risks whose likelihood cannot be calculated, it is appropriate to act, not to stand by and merely hope.

Writing in a similar vein before "precaution" became widely discussed, Wildavsky (1979) argued that some risk taking is essential to social progress. Thompson (1980) has made very similar arguments in comparing societies and cultures.

Precaution is often in the eye of the beholder. Thus, for example, some have argued that while the European Union has been more precautionary with respect to climate change and CO_2 emissions in promoting the wide adoption of fuel efficient diesel automobiles, the United States has been more precautionary with respect to health effects of fine particulate air pollution, stalling the adoption of diesel automobiles until it was possible to substantially reduce their particulate emissions (Wiener and Rogers, 2002).

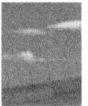

Communicating Uncertainty

Lead Author: M. Granger Morgan, Carnegie Mellon Univ.

Contributing Authors: Hadi Dowlatabadi, Univ. of British Columbia; Max Henrion, Lumina Decision Systems; David Keith, Univ. of Calgary; Robert Lempert, The RAND Corporation; Sandra McBride, Duke Univ.; Mitchell Small, Carnegie Mellon Univ.; Thomas Wilbanks, Oak Ridge National Laboratory

It is often argued that one should not try to communicate about uncertainty to non-technical audiences[1], because laypeople won't understand and decision makers want definitive answers—what Senator Muskie referred to as the ideal of receiving advice from "one armed scientists"[2].

We do not agree. Non-technical people deal with uncertainty and statements of probability all the time. They don't always reason correctly about probability, but they can generally get the gist (Dawes, 1988). While they may make errors about the details, for the most part they manage to deal with probabilistic weather forecasts about the likelihood of rain or snow, point spreads at the track, and similar probabilistic information. The real issue is to frame things in familiar and understandable terms[3].

There has been considerable discussion in the literature about whether it is best to present uncertainties to laypeople in terms of odds (*e.g.*, 1 in 1,000) or probabilities (*e.g.*, p = 0.001)[4] (Fischhoff *et al.*, 2002). Baruch Fischhoff provides the following summary advice:

- Either will work, if they're used consistently across many presentations.
- If you want people to understand one fact, in isolation, present the result both in terms of odds and probabilities.
- In many cases, there's probably more confusion about what is meant by the specific events being discussed than about the numbers attached to them.

---

[1] By "non-technical audiences" we mean people who have not had courses or other serious exposure to the basic ideas of science past the level of high school.

[2] The reference, of course, being to experts who always answered his questions "on the one hand…but on the other hand…," the phrase is usually first attributed to Senator Edmund Muskie.

[3] Several of the statements in this paragraph are consistent with the findings of a workshop run by the NOAA Office of the Federal Coordinator for Meteorological (OFCM, 2001).

[4] Strictly odds are defined as p/(1-p) but when p is small, for simplicity the difference between odds of 1 in 999 and 1 in 1,000 is often ignored when presenting results to non-technical audiences.

"People have a tendency to overestimate the likelihood of low-magnitude events, and under-estimate the likelihood of high-magnitude events".

Ibrekk and Morgan (1987) reached a similar conclusion in their study of alternative simple graphical displays for communicating uncertainty to non-technical people, arguing for the use of more than one display when communicating a single uncertain result. They also report that "rusty or limited statistical knowledge does not significantly improve the performance of semi-technical or laypersons in interpreting displays that communicate uncertainty" (Morgan and Henrion, 1990).

Patt and Schrag (2003) studied how undergraduate respondents interpret both probabilities and uncertainty words that specifically relate to climate and weather. They found that these respondents mediated their probability judgments by the severity of the event reported (*e.g.*, hurricane *versus* snow flurries). They conclude that "in response to a fixed probability scale, people will have a tendency to overestimate the likelihood of low-magnitude events, and under-estimate the likelihood of high-magnitude events". This is because "intuitively people use such language to describe both the probability and the magnitude of risks, and they expect communicators to do the same". They suggest that unless analysts make it clear that they are not adjusting their probability estimates up and down depending on the severity of the event described, policy makers' response to assessments are "...likely to be biased downward, leading to insufficient efforts to mitigate and adapt to climate change".

The presence of high levels of uncertainty offers people with an agenda an opportunity to "spin the facts". Dowlatabadi reports that when he first started showing probabilistic outputs from Carnegie Mellon's Integrated Climate Assessment Model (ICAM) to staff on Capitol Hill, many of those who thought that climate change was not happening or was not important, immediately focused in on the low impact ends of the model's probabilistic outputs. In contrast, many of those who thought climate change was a very serious problem immediately focused in on the high impact ends of the model's probabilistic outputs.

This does not mean that one should abandon communicating about uncertainty. There will always be people who wish to distort the truth.

However, it does mean that communicating uncertainty in key issues requires special care, so that those who really want to understand can do so.

Recipients will process any message they receive through their previous knowledge and perception of the issues at hand. Thus, in designing an effective communication, one must first understand what folks who will receive that message already know and think about the topics at hand. One of the clearest findings in the empirical literature on risk communication is that no one can design effective risk communication messages without some empirical evaluation and refinement of those messages with members of the target audience.

In order to support the design of effective risk communication messages, Morgan *et al.* (2002) and colleagues developed a "mental model" approach to risk communication. Using open-ended interview methods, subjects are asked to talk about the issues at hand, with the interviewer providing as little structure or input to the interview process as possible. After a modest number of interviews have been conducted, typically 20 or so, an asymptote is reached in the concepts mentioned by the interviewees and few additional concepts are encountered. Once a set of key issues and perceptions have been identified, a closed form survey is developed that can be used to examine which of the concepts are most prevalent, and which are simply the idiosyncratic response of a single respondent. The importance of continued and iterative empirical evaluation of the effectiveness of communication is stressed.

One key finding is that empirical study is absolutely essential to the development of effective communication. With this in mind, there is no such thing as an expert in communication—in the sense of someone who can tell you ahead of time (*i.e.*, without empirical study) how a message should be framed, or what it should say.

Using this method, Bostrom *et al.* (1994) and Read *et al.* (1994) examined public understanding and perception of climate change. On the basis of their findings, a communication brochure for the general public was developed and iteratively refined using read-aloud protocols and focus group discussions (Morgan and Smuts, 1994). Using less formal ethnographic methods, Kempton (1991) and Kempton *et al.* (1995) have conducted studies of public perceptions of climate change and related issues, obtaining results that are very similar to those of the mental model studies. More recently, Reiner *et al.* (2006) have conducted a cross-national study of some similar issues.

While the preceding discussion has dealt with communicating uncertainty in situations in which it is possible to do extensive studies of the relative effectiveness of different communication methods and messages, much of the communication about uncertain events that all of us receive comes from reading or listening to the press.

Philip M. Boffey (quoted in Friedman *et al.*, 1999), editorial page editor for *The New York Times*, argues that "uncertainty is a smaller problem for science writers than for many other kinds of journalists". He notes that there is enormous uncertainty about what is going on in China or North Korea and that "economics is another area where there is great uncertainty". In contrast, he notes:

> With science writing, the subjects are better defined. One of the reasons why uncertainty is less of a problem for a science journalist is because the scientific material we cover is mostly issued and argued publicly. This is not North Korea or China. While it is true that a journalist cannot view a scientist's lab notes or sit on a peer review committee, the final product is out there in the public. There can be a vigorous public debate about it and reporters and others can see what is happening.

Boffey goes on to note that "one of the problems in journalism is to try to find out what is really happening". While this may be easier than in some other fields, because of peer-reviewed articles, consensus panel mechanisms such as National Research Council (NRC) reports, "there is the second level problem of deciding whether these consensus mechanisms are operating properly…Often the journalist does not have time to investigate…given the constraints of daily journalism". However, he notes:

> …these consensus mechanisms do help the journalist decide where the mainstream opinion is and how and whether to deal with outliers. Should they be part of the debate? In some issues, such as climate change, I do not feel they should be ignored because in this subject, the last major consensus report showed that there were a number of unknowns, so the situation is still fluid….

While it is by no means unique, climate change is perhaps the prototypical example of an issue for which there is a combination of considerable scientific uncertainty, and strong short-term economic and other interests at play. Uncertainty offers the opportunity for various interests to confuse and divert the public discourse in what may already be a very difficult scientific process of seeking improved insight and understanding. In addition, many reporters are not in a position to make their own independent assessment of the likely accuracy of scientific statements. They have a tendency to seek conflict and report "on the one hand, on the other hand", doing so in just a few words and with very short deadlines. It is small wonder that sometimes there are problems.

Chemist and Nobel laureate Sherwood Roland (quoted in Friedman *et al.*, 1999) notes that "…scientists' reputations depend on their findings being right most of the time. Sometimes, however, there are people who are wrong almost all the time and they are still quoted in the media 20 years later very consistently".

Despite continued discourse within scientific societies and similar professional circles about the importance of scientists interpreting and communicating their findings to the public and to decision makers, freelance environmental writer Dianne Dumanoski (quoted in Friedman *et al.*, 1999) observes that "strong peer pressure exists within the scientific community against becoming a visible scientist who communicates with the media and the public". This pressure, combined with an environment in which there is high probability that many statements a scientist makes about uncertainties will immediately be seized upon by advocates in an ongoing public debate, it is perhaps understandable that many scientists choose to just keep their heads down,

Uncertainty offers the opportunity for various interests to confuse and divert the public discourse in what may already be a very difficult scientific process of seeking improved insight and understanding.

do their research, and limit their communication to publication in scientific journals and presentations at professional scientific meetings.

The problems are well illustrated in an exchange between biological scientist Rita Colwell (then Director of the National Science Foundation), Peggy Girsham of NBC (now with NPR) and Sherry Roland reported by Friedman *et al.* (1999). Colwell noted that when a scientist talks with a reporter, they must be very careful about what they say, especially if they have a theory or findings that run counter to conventional scientific wisdom. She observed that "it is very tough to go out there, talk to a reporter, lay your reputation on the line and then be maligned by so called authorities in a very unpleasant way". She noted that this problem is particularly true for women scientists, adding "I have literally taken slander and public ridicule from a few individuals with clout and that has been very unpleasant...." NBC's Girsham (now with NPR) noted that, in a way, scientists in such a situation cannot win "because if you are not willing to talk to a reporter, then we [in the press] will look for someone who is willing and may be less cautious about expressing a point of view". Building on this point, Rowland noted that in the early days of the work he and Mario Molina did on stratospheric ozone depletion, "Molina and I read *Aerosol Age* avidly because we were the 'black hats' in every issue. The magazine even went so far as to run an article calling us agents of the Soviet Union's KGB, who were trying to destroy American industry...what was more disturbing was when scientists on the industry side were quoted by the media, claiming our calculations of how many CFCs were in the stratosphere were off by a factor of 1,000...even after we won the Nobel Prize for this research, our politically conservative local newspaper... [said that while the] theory had been demonstrated in the laboratory... scientists with more expertise in atmospheric science had shown that the evidence in the real atmosphere was quite mixed. This ignored the consensus views of the world's atmospheric scientists that the results had been spectacularly confirmed in the real atmosphere". Clearly, even when a scientist is as careful and balanced as possible, communicating with the public and decisions makers about complex and politically contentious scientific issues is not for the faint hearted!

Some Simple Guidance for Researchers

Lead Author: M. Granger Morgan, Carnegie Mellon Univ.

Contributing Authors: Hadi Dowlatabadi, Univ. of British
Columbia; Max Henrion, Lumina Decision Systems; David Keith,
Univ. of Calgary; Robert Lempert, The RAND Corporation; Sandra
McBride, Duke Univ.; Mitchell Small, Carnegie Mellon Univ.; Thomas
Wilbanks, Oak Ridge National Laboratory

Doing a good job of characterizing and dealing with uncertainty can never be reduced to a simple cookbook. One must always think critically and continually ask questions such as:

- Does what we are doing make sense?
- Are there other important factors that are equally or more important than the factors we are considering?
- Are there key correlation structures in the problem that are being ignored?
- Are there normative assumptions and judgments about which we are not being explicit?
- Is information about the uncertainties related to research results and potential policies being communicated clearly and consistently?

That said, the following are a few words of guidance to help Climate Change Science Program (CCSP) researchers and analysts to do a better job of reporting, characterizing, and analyzing uncertainty[1]. Some of this guidance is based on available literature. However, because doing these things well is often as much an art as it is a science, the recommendations also draw on the very considerable[2] and diverse experience and collective judgment of the writing team.

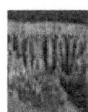

---

[1] The advice in this Part is intended for use by analysts addressing a range of climate problems in the future. For a variety of reasons, many of the CCSP products have already been produced and obviously will not be able to follow advice provided in this Part. Most others are well along in production and thus will also not be able to adopt advice provided here. However, the current round of CCSP products is certainly not the last word in the analysis or assessment of climate change, its impacts, or in the development of strategies and policies for abatement and adaptation.

[2] Collectively the author team has roughly 200 person-years of experience in addressing these issues both theoretically and in practical analysis in the context of climate and other similar areas.

Associating Common Language with Subjective Probability

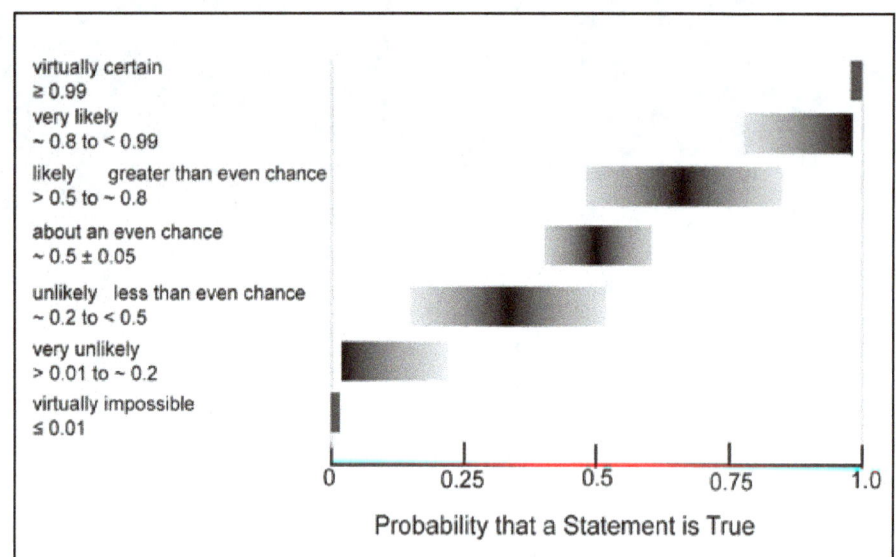

Figure 9.1 Recommended framework for associating common language with subjective probability values.

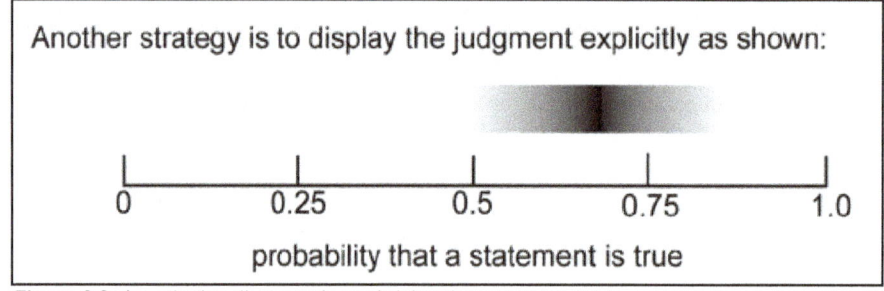

Figure 9.2 A method to illustrate the probability that a statement is true.

Recommended Box Plot Format

Figure 9.3 Recommended format for box plots. When many uncertain results are to be reported, box plots can be stacked more compactly than probability distributions.

Reporting Uncertainty

- When qualitative uncertainty words such as "likely" and "unlikely" are used, it is important to clarify the range of subjective probability values that are to be associated with those words. Unless there is some compelling reason to do otherwise, we recommend the use of the framework shown in figure 9.1[3]:

Another strategy is to display the judgment explicitly as shown in figure 9.2:

This approach provides somewhat greater precision and allows some limited indication of secondary uncertainty for those who feel uncomfortable making precise probability judgments.

- In any document that reports uncertainties in conventional scientific format (*e.g.*, 3.5±0.7), it is important to be explicit about what uncertainty is being included and what is not, and to explain what range is being reported (*e.g.*, plus or minus one standard error of the mean, two standard deviations, *etc.*). This reporting format is generally not appropriate for large uncertainties or where distributions have a lower or upper bound and hence are not symmetric. In all cases, care should be taken not to report results using more significant figures than are warranted by the associated uncertainty. Often this means overriding default values on standard software such as Microsoft Excel.
- Care should be taken in plotting and labeling the vertical axes when reporting prob-

ability density functions (PDFs). The units are probability density (*i.e.*, probability per unit interval along the horizontal axis), not probability.

- Since many people find it difficult to read and correctly interpret PDFs and cumulative distribution functions (CDFs), when space allows, it is best practice to plot the CDF together with the PDF on the same x-axis (Morgan and Henrion, 1990).
- While it is always best to report results in terms of full PDFs and/or CDFs, when many uncertain results must be reported, box plots (first popularized by Tukey, 1977) are often the best way to do this in a compact manner. There are several conventions. Our recommendation is shown in figure 9.3, but what is most important is to be clear about the notation.
- While there may be a few circumstances in which it is desirable or necessary to address and deal with second-order uncertainty (*e.g.*, how sure an expert is about the shape of an elicited CDF), more often than not the desire to perform such analysis arises from a misunderstanding of the nature of subjective probabilistic statements (see the discussion in Part 1). When second-order uncertainty is being considered, one should be very careful to determine that the added level of such complication will aid in, and will not unnecessarily complicate, subsequent use of the results.

Characterizing and Analyzing Uncertainty

Unless there are compelling reasons to do otherwise, conventional probability is the best tool for characterizing and analyzing uncertainty about climate change and its impacts.

- The elicitation of expert judgment, often in the form of subjective probability distributions, can be a useful way to combine the formal knowledge in a field as reflected in the literature with the informal knowledge and physical intuition of experts. Elicitation is not a substitute for doing the needed science, but it can be a very useful tool in support of research planning, private decision making, and the formulation of public policy.

---

[3] This display divides the interval between 0.99 and 0.01 into five ranges, adding somewhat more resolution across this range than the mapping used by IPCC (2001). However, it is far more important to map words into probabilities in a consistent way, *and to be explicit about how that is being done*, than it is to use any specific mapping. Words are inherently imprecise. In the draft version of this diagram, we intentionally included significantly greater overlap between the categories. A number of reviewers were uncomfortable with this overlap, calling for a precise one-to-one mapping between words and probabilities. On the other hand, when a draft of the United States National Assessment (NAST, 2001) produced a diagram with such a precise mapping, reviewers complained about the precise boundaries, with the result that in the final version they were made fuzzy (Figure 2.3). For a more extended discussion of these issues, see Part 2 of this Product.

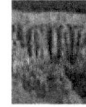

However, the design and execution of a good expert elicitation takes time and requires a careful integration of knowledge of the relevant substantive domain with knowledge of behavioral decision science (see discussion in Part 5).

- When eliciting probability distributions from multiple experts, if they disagree significantly, it is generally better to report the distributions separately. This is especially true if such judgments will subsequently be used as inputs to a model that has a non-linear response.

- There are a variety of software tools available to support probabilistic analysis using Monte Carlo and related techniques. As with any powerful analytical tool, their proper use requires careful thought and care.

- In performing uncertainty analysis, it is important to think carefully about possible sources of correlation. One simple procedure for getting a sense of how important this may be is to run the analysis with key variables uncorrelated and then run it again with key variables perfectly correlated. Often, in answering questions about aggregate parameter values, experts assume correlation structures between the various components of the aggregate value being elicited. Sometimes it is important to elicit the component uncertainties separately from the aggregate uncertainty in order to reason out why specific correlation structures are being assumed.

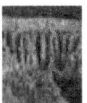

- Methods for describing and dealing with data pedigree (*e.g.*, Funtowicz and Ravetz, 1990) have not been developed to the point that they can be effectively incorporated in probabilistic analysis. However, the quality of the data on which judgments are based is clearly important and should be addressed, especially when uncertain information of varying quality and reliability is combined in a single analysis. At a minimum, investigators should be careful to provide a "traceable account" of where their results and judgments have come from.

- While full probabilistic analysis can be useful, in many contexts, simple parametric analysis, or back-to-front analysis (that works backwards from an end point of interest) may be as or more effective in identifying key unknowns and critical

levels of knowledge needed to make better decisions.

- Scenarios analysis can be useful, but also carries risks. Specific detailed scenarios can become cognitively compelling, with the result that people may overlook many other pathways to the same end-points. It is often best to "cut the long causal chains" and focus on the possible range of a few key variables, which can most affect outcomes of interest.

- Scenarios, which describe a single point (or line) in a multi-dimensional space, cannot be assigned probabilities. If, as is often the case, it will be useful to assign probabilities to scenarios, they should be defined in terms of intervals in the space of interest, not in terms of point values.

- Variability and uncertainty is not the same thing. Sometimes it is important to draw distinction between the two but often it is not. A distinction should be made only when it adds clarity for users.

- Analysis that yields predictions is very helpful when our knowledge is sufficient to make meaningful predictions. However, the past history of success in such efforts suggests great caution (*e.g.*, Chapters 3 and 6 in Smil, 2003). When meaningful prediction is not possible, alternative strategies, such as searching for responses or policies that will be robust across a wide range of possible futures, deserve careful consideration.

For some problems there comes a time when uncertainty is so high that conventional modes of probabilistic analysis (including decision analysis) may no longer make sense. While it is not easy to identify this point, investigators should continually ask themselves whether what they are doing makes sense and whether a much simpler approach, such as a bounding or order-of-magnitude analysis, might be superior (*e.g.*, Casman *et al.*, 1999).

GLOSSARY AND ACRONYMS

ACRONYMS AND ABBREVIATIONS

3-D:	three dimensional
AIC:	Akaike's Information Criterion
AMOC:	Atlantic meridional overturning circulation
AR4:	Fourth Assessment Report (IPCC)
ARMA:	autoregressive moving average
BIC:	Bayesian information criterion
CCN:	cloud condensation nuclei
CCSP:	Climate Change Science Program
CDF:	cumulative density function
CEI:	Climate Extremes Index
CFC:	chlorofluorocarbon
CO_2:	carbon dioxide
ENSO:	El Niño Southern Oscillation
EPA:	Environmental Protection Agency
GCM:	general circulation model
GCRI:	Greenhouse Climate Response Index
GHG:	greenhouse gas
HadAM2:	Hadley Centre Atmosphere Model version 2 (from the Hadley Centre for Climate Prediction and Research, United Kingdom Meteorological Office)
HadSM3:	Hadley Centre Slab Climate Model version 3 (from the Hadley Centre for Climate Prediction and Research, United Kingdom Meteorological Office)
ICAM:	Integrated Climate Assessment Model
IGSM:	Integrated Global System Model
IPCC:	Intergovernmental Panel on Climate Change
K:	kelvin (temperature)
LR:	likelihood ratio
MCA:	maximum covariance analysis
MIT:	Massachusetts Institute of Technology
NAO:	North Atlantic oscillation
NO_x:	nitrogen oxides
NRC:	National Research Council
PCA:	principal component analysis
PDF:	probability density function
PPE:	perturbed physics ensemble
PPM:	parts per million
PPMV:	parts per million by volume
PROPS:	predictive oscillation patterns
SAP:	Synthesis and Assessment Product
SO_x:	sulfur oxides
SO_2:	sulfur dioxide
TAR:	Third Assessment Report (IPCC)
TFI:	technical facilitator/integrator
WG:	Working Group (IPCC)

REFERENCES

NON-TECHNICAL SUMMARY REFERENCES

Dawes, R.M., 1988: *Rational Choice in an Uncertain World.* Harcourt Brace Jovanovich, San Diego, CA, 346 pp.

Fischhoff, B., A. Bostrom, and M. Jacobs-Quadrel, 2002: Risk perception and communication. In: *Oxford Textbook of Public Health* [Detels, R., J. McEwen, R. Reaglenhole, and H. Tanaka (eds.)]. Oxford University Press, New York, 4th ed., pp. 1105-1123.

Franklin, B., 1789: Letter to Jean-Baptiste Leroy.

Rumsfeld, D., 2002: News briefing as quoted by M. Shermer, 2005: Rumsfeld's wisdom. *Scientific American*, **293(3)**, 38.

Smil, V., 2007: *Global Catastrophes and Trends: The Next Fifty Years.* MIT Press, Cambridge, MA, 307 pp.

PART 1 REFERENCES

Cyert, R.M. and M.H. DeGroot, 1987: *Bayesian Analysis and Uncertainty in Economic Theory.* Rowman and Littlefield, Totowa, NJ, 206 pp.

Devaney, R.L., 2003: *An Introduction to Chaotic Dynamical Systems.* Westview Press, Boulder, CO, 2nd edition, 335 pp.

Dumanoski, D. (quoted in Friedman *et al.*, 1999).

Ellsberg, D., 1961: Risk, ambiguity and the savage axioms. *Quarterly Journal of Economics*, **75(4)**, 643-669.

Friedman, S.M., S. Dunwoody, and C.L. Rogers, 1999: *Communicating Uncertainty: Media Coverage of New and Controversial Science.* L. Erlbaum, Mahwah, NJ, 277 pp.

Good, I.J., 1962: How rational should a manager be? *Management Science*, **8(4)**, 383-393.

Henrion, M., 1999: Uncertainty. In: *MIT Encyclopedia of the Cognitive Sciences* [Wilson, R.A. and F.C. Keil (eds.)]. The MIT Press, Cambridge, MA.

IPCC (Intergovernmental Panel on Climate Change), 2005: *Guidance Notes for Lead Authors of the IPCC Fourth Assessment Report on Addressing Uncertainties.* [Intergovernmental Panel on Climate Change, Geneva, Switzerland], 4 pp. Available at: http://www.ipcc.ch/pdf/supporting-material/uncertainty-guidance-note.pdf

Knight, F.H., 1921: *Risk, Uncertainty and Profit.* Houghton Mifflin Company, Boston, MA, 381 pp.

Kuhn, T.S., 1962: *The Structure of Scientific Revolutions.* University of Chicago Press, Chicago, IL, 172 pp.

Lorenz, E.N., 1963: Deterministic nonperiodic flow. *Journal of the Atmospheric Sciences*, **20(2)**, 130-141.

Morgan, M.G. and M. Henrion, 1990: *Uncertainty: A Guide to Dealing with Uncertainty in Quantitative Risk and Policy Analysis.* Cambridge University Press, Cambridge, UK, and New York, 332 pp.

Moss, R. and S.H. Schneider, 2000: Uncertainties in the IPCC TAR: Recommendations to lead authors for more consistent assessment and reporting. In: *Guidance Papers on the Cross Cutting Issues of the Third Assessment Report of the IPCC* [Pachauri, R., T. Taniguchi, and K. Tanaka (eds.)]. Intergovernmental Panel on Climate Change, Geneva, Switzerland, pp. 33-51. Available at: http://www.ipcc.ch/pdf/supporting-material/guidance-papers-3rd-assessment.pdf

Paté-Cornell, M.E., 1996: Uncertainties in risk analysis: Six levels of treatment. *Reliability Engineering and System Safety*, **54(2-3)**, 95-111.

Rumsfeld, D., 2002: News briefing as quoted by M. Shermer, 2005: Rumsfeld's wisdom. *Scientific American*, **293(3)**, 38.

Smithson, M., 1988: *Ignorance and Uncertainty: Emerging Paradigms.* Springer-Verlag, New York, 393 pp.

Weick, K.E. and K.M. Sutcliffe, 2001: *Managing the Unexpected: Assuring High Performance in an Age of Complexity.* Jossey-Bass, San Francisco, CA, 200 pp.

PART 2 REFERENCES

Clark, H.H., 1990: [Quantifying probabilistic expressions] Comment. *Statistical Science*, **5(1)**, 12-16.

Cliff, N., 1990: [Quantifying probabilistic expressions] Comment, *Statistical Science*, **5(1)**, 16-18.

IPCC (Intergovernmental Panel on Climate Change), 2001a: *Climate Change 2001: The Scientific Basis.* Contribution of Working Group I to the Third Assessment Report of the Intergovernmental Panel on Climate Change [Houghton, J.T., Y. Ding, D.J. Griggs, M. Noguer, P.J. van der Linden, X.

Dai, K. Maskell, and C.A. Johnson (eds.)]. Cambridge University Press, Cambridge, UK, and New York, 881 pp.

IPCC (Intergovernmental Panel on Climate Change), 2001b: *Climate Change 2001: Impacts, Adaptation, and Vulnerability*. Contribution of Working Group II to the Third Assessment Report of the Intergovernmental Panel on Climate Change [McCarthy, J.J., O.F. Canziani, N.A. Leary, D.J. Dokken, and K.S. White (eds.)]. Cambridge University Press, Cambridge, UK, and New York, 1032 pp.

IPCC (Intergovernmental Panel on Climate Change), 2004: *Workshop on Describing Scientific Uncertainties in Climate Change to Support Analysis of Risk and of Options*, May 11-13, 2004, National University of Ireland, Maynooth, Co. Kildare, Ireland. [Manning, M., M. Petit, D. Easterling, J. Murphy, A. Patwardhan, H.-H. Rogner, R. Swart, and G. Yohe (eds.)]. Working Group I Technical Support Unit, Boulder, Colorado, 146 pp. Available at: http://www.ipcc.ch/pdf/supporting-material/ ipcc-workshop-2004-may.pdf

IPCC (Intergovernmental Panel on Climate Change), 2005: *Guidance Notes for Lead Authors of the IPCC Fourth Assessment Report on Addressing Uncertainties*. [Intergovernmental Panel on Climate Change, Geneva, Switzerland], 4 pp. Available at: http://www.ipcc.ch/pdf/supporting-material/uncertainty-guidance-note.pdf

IPCC (Intergovernmental Panel on Climate Change), 2007: *Climate Change 2007: The Physical Science Basis*. Contribution of Working Group I to the Fourth Assessment Report of the Intergovernmental Panel on Climate Change [Solomon, S., D. Qin, M. Manning, Z. Chen, M. Marquis, K. Averyt, M.M.B. Tignor, and H.L. Miller (eds.)]. Cambridge University Press, Cambridge, UK, and New York, 996 pp.

Kadane, J.B., 1990: [Quantifying probabilistic expressions] Comment: Codifying chance. *Statistical Science*, **5(1)**, 18-20.

Kruskal, W., 1990: [Quantifying probabilistic expressions] Comment. *Statistical Science*, **5(1)**, 20-21.

Morgan, M.G., 1998: Uncertainty analysis in risk assessment. *Human and Ecological Risk Assessment*, **4(1)**, 25-39.

Moss, R. and S.H. Schneider, 2000: Uncertainties in the IPCC TAR: Recommendations to lead authors for more consistent assessment and reporting. In: *Guidance Papers on the Cross Cutting Issues of the Third Assessment Report of the IPCC* [Pachauri, R., T. Taniguchi, and K. Tanaka (eds.)]. Intergovernmental Panel on Climate Change, Geneva, Switzerland, pp. 33-51. Available at: http://www.ipcc.ch/pdf/supporting-material/guidance-papers-3rd-assessment.pdf

Mosteller, F. and C. Youtz, 1990: Quantifying probabilistic expressions. *Statistical Science*, **5(1)**, 2-12.

NAST (National Assessment Synthesis Team), 2001: *Climate Change Impacts on the United States: The Potential Consequences of Climate Variability and Change*. Cambridge University Press, Cambridge, UK, and New York, 612 pp.

Presidential/Congressional Commission on Risk Assessment and Risk Management, 1997: *Final Report: Volume 1: Framework for Environmental Health Risk Management; Volume 2: Risk Assessment and Risk Management in Regulatory Decision-Making*. The Commission, Washington, DC.

Tanur, J.M., 1990: [Quantifying probabilistic expressions] Comment: On the possible dangers of isolation. *Statistical Science*, **5(1)**, 21-22.

U.S. EPA (United States Environmental Protection Agency), 1996: *Proposed Guidelines for Carcinogenic Risk Assessment*. EPA/600/P-92/003C. Office of Research and Development, Environmental Protection Agency, Washington DC, 170 pp.

Wallsten, T.S. and D.V. Budescu, 1990: [Quantifying probabilistic expressions] Comment. *Statistical Science*, **5(1)**, 23-26.

Wallsten, T.S., D.V. Budescu, A. Rapoport, R. Zwick, and B. Forsyth, 1986: Measuring the vague meanings of probability terms. *Journal of Experimental Psychology: General*, **155(4)**, 348-365.

Winkler, R.L., 1990: [Quantifying probabilistic expressions] Comment: Representing and communicating uncertainty. *Statistical Science*, **5(1)**, 26-30.

Wolf, C. Jr., 1990: [Quantifying probabilistic expressions] Comment. *Statistical Science*, **5(1)**, 31-32.

PART 3 REFERENCES

DeGroot, M., 1965: *Thought and Choice in Chess*. Basic Books, New York, 463 pp.

Hastie, R. and R.M. Dawes, 2001: *Rational Choice in an Uncertain World: The Psychology of Judgment and Decision Making*. Sage, Thousand Oaks, CA, 372 pp.

Henrion, M. and B. Fischhoff, 1986: Assessing uncertainty in physical constants. *American Journal of Physics*, **54(9)**, 791-798.

Kahneman, D., P. Slovic, and A. Tversky (eds.), 1982: *Judgment Under Uncertainty: Heuristics and Biases*. Cambridge University Press, Cambridge, UK, and New York, 551 pp.

Lichtenstein, S., P. Slovic, B. Fischhoff, M. Layman, and B. Combs, 1978: Judged frequency of lethal events. *Journal of Experimental Psychology: Human Learning and Memory*, **4(6)**, 551-578.

Lichtenstein, S., B. Fischhoff, and L.D. Phillips, 1982: Calibration of probabilities: The state of the art to 1980. In: *Judgment Under Uncertainty: Heuristics and Biases* [Kahneman, D., P. Slovic, and A. Tversky (eds.)]. Cambridge University Press, Cambridge, UK, and New York, pp. 306-334.

Loewenstein, G.F., 1996: Out of control: Visceral influences on behavior. *Organizational Behavior and Human Decision Processes*, **65(3)**, 272-292.

Loewenstein, G.F., E.U. Weber, C.K. Hsee, and E.S. Welch, 2001: Risk as feelings. *Psychological Bulletin*, **127(2)**, 267-286.

Morgan, M.G. and M. Henrion, 1990: *Uncertainty: A Guide to Dealing with Uncertainty in Quantitative Risk and Policy Analysis*. Cambridge University Press, Cambridge, UK, and New York, 332 pp.

Sjöberg, L., 2006: Will the real meaning of affect please stand up? *Journal of Risk Research*, **9(2)**, 101-108.

Slovic, P., M.L. Finucane, E. Peters, and D.G. MacGregor, 2004: Risk as analysis and risk as feelings: Some thoughts about affect, reason, risk and rationality. *Risk Analysis*, **24(2)**, 311-322.

Tversky, A. and D. Kahneman, 1974: Judgments under uncertainty: Heuristics and biases. *Science*, **185(4157)**, 1124-1131.

Wardman, J.K., 2006: Toward a critical discourse on affect and risk perception. *Journal of Risk Research*, **9(2)**, 109-124.

PART 4 REFERENCES

Allen, M.R. and W.J. Ingram, 2002: Constraints on the future changes in climate and the hydrological cycle. *Nature*, **419(6904)**, 224-232.

Andronova, N. and M.E. Schlesinger, 2001: Objective estimation of the probability distribution for climate sensitivity. *Journal of Geophysical Research*, **106(D19)**, 22605-22612.

Annan, J.D., 2005: Parameter estimation using chaotic time series. *Tellus A*, **57(5)**, 709-714.

Annan, J.D., J.C. Hargreaves, N.R. Edwards, and R. Marsh, 2005: Parameter estimation in an intermediate complexity earth system model using an ensemble Kalman filter. *Ocean Modelling*, **8(1-2)**, 135-154.

Austin, M.P., 2002: Spatial prediction of species distribution: An interface between ecological theory and statistical modeling. *Ecological Modelling*, **157(2)**, 101-118.

Bell, J., P.B. Duffy, C. Covey, L. Sloan, and the CMIP Investigators, 2000: Comparison of temperature variability in observa-

tions and sixteen climate model simulations. *Geophysical Research Letters*, **27(2)**, 261-264.

Berger, J.O., 1994: An overview of robust Bayesian analysis (with discussion). *Test*, **3(1)**, 5-124.

Berliner, L.M., 2003: Physical-statistical modeling in geophysics. *Journal of Geophysical Research*, **108(D24)**, 8776, doi:10.1029/2002JD002865.

Berliner, L.M., J.A. Royle, C.K. Wilke, and R.F. Milliff, 1999: Bayesian methods in the atmospheric sciences. In: *Bayesian Statistics 6* [Bernardo, J.M., J.O. Berger, A.P. Dawid and A.F.M. Smith (eds.)]. Oxford University Press, New York, pp. 83-100.

Berliner, L.M., R.A. Levine, and D.J. Shea, 2000: Bayesian climate change assessment. *Journal of Climate*, **13(21)**, 3805-3820.

Charles, S.P., B.C. Bates, I.N. Smith, and J.P. Hughes, 2004: Statistical downscaling of daily precipitation from observed and modeled atmospheric fields. *Hydrological Processes*, **18(8)**, 1373-1394.

Daley, R., 1997: Atmospheric data assimilation. *Journal of the Meteorological Society of Japan*, **75(1B)**, 319-329.

Edwards, N.R. and R. Marsh, 2005: Uncertainties due to transport-parameter sensitivity in an efficient 3D ocean-climate model. *Climate Dynamics*, **24(4)**, 415-433.

Epstein, E.S., 1985: *Statistical Inference and Prediction in Climatology: A Bayesian Approach*. Meteorological monographs v. 20 no. 42. American Meteorological Society, Boston, MA, 199 pp.

Evensen, G. and P.J. van Leeuwen, 2000: An ensemble Kalman smoother for nonlinear dynamics. *Monthly Weather Review*, **128(6)**, 1852-1867.

Fuentes, M., P. Guttorp, and P. Challenor, 2003: Statistical assessment of numerical models. *International Statistical Review*, **71(2)**, 201-221.

Goldstein, M., 2006: Subjective Bayesian analysis: Principles and practice. *Bayesian Analysis*, **1(3)**, 403-420.

Goldstein, M. and J. Rougier, 2006: Bayes linear calibrated prediction for complex systems. *Journal of the American Statistical Association*, **101(475)**, 1132-1143.

Hasselmann, K., 1979: On the signal-to-noise problem in atmospheric response studies. In: *Meteorology of Tropical Oceans* [D.B. Shaw (ed.)]. Royal Meteorological Society, Bracknell, Berkshire (UK), pp. 251-259.

Hasselmann, K., 1993: Optimal fingerprints for the detection of time-dependent climate change. *Journal of Climate*, **6(10)**, 1957-1971.

Hughes, J.P., P. Guttorp, and S.P. Charles, 1999: A non-homogeneous hidden Markov model for precipitation occurrence. *Journal of the Royal Statistical Society: Series C (Applied Statistics)*, **48(1)**, 15-30.

Jones, C.D., P.M. Cox, and C. Huntingford, 2006: Climate-carbon cycle feedbacks under stabilization: Uncertainty and observational constraints. *Tellus B*, **58(5)**, 603-613.

Kallache, M., H.W. Rust, and J. Kropp, 2005: Trend assessment: Applications for hydrology and climate research. *Nonlinear Processes in Geophysics*, **12(2)**, 201-210.

Karl, T.R., R.W. Knight, D.R. Easterling, and R.G. Quayle, 1996: Indices of climate change for the United States. *Bulletin of the American Meteorological Society*, **77(2)**, 279-292.

Katz, R.W., 2002: Techniques for estimating uncertainty in climate change scenarios and impact studies. *Climate Research*, **20(2)**, 167-185.

Katz, R.W. and M. Ehrendorfer, 2006: Bayesian approach to decision making using ensemble weather forecasts. *Weather and Forecasting*, **21(2)**, 220-231.

Kennedy, M.C. and A. O'Hagan, 2001: Bayesian calibration of computer models. *Journal of the Royal Statistical Society, Series B*, **63(3)**, 425-464.

Kennedy, M.C., C.W. Anderson, S. Conti, and A. O'Hagan, 2006: Case studies in Gaussian process modelling of computer codes. *Reliability Engineering and System Safety*, **91(10-11)**, 1301-1309.

Kheshgi, H.S. and B.S. White, 2001: Testing distributed parameter hypotheses for the detection of climate change. *Journal Climate*, **14(16)**, 3464-3481.

Kooperberg, C. and F. O'Sullivan, 1996: Predictive oscillation patterns: A synthesis of methods for spatial-temporal decomposition of random fields. *Journal of the American Statistical Association*, **91(436)**, 1485-1496.

Koutsoyiannis, D., 2003: Climate change, the Hurst phenomenon, and hydrological statistics. *Hydrological Sciences Journal*, **48(1)**, 3-24.

Laud, P.W. and J.G. Ibrahim, 1995: Predictive model selection. *Journal of the Royal Statistical Society, Series B*, **57(1)**, 247-262.

Levine, R.A. and L.M. Berliner, 1999: Statistical principles for climate change studies. *Journal of Climate*, **12(2)**, 564-574.

Lund, R. and J. Reeves, 2002: Detection of undocumented change-points: A revision of the two-phase regression model. *Journal of Climate*, **15(17)**, 2547-2554.

McAvaney, B.J., C. Covey, S. Joussaume, V. Kattsov, A. Kitoh, W. Ogana, A.J. Pitman, A.J. Weaver, R.A. Wood, and Z.C. Zhao, 2001: Model evaluation. In: *Climate Change 2001: The Scientific Basis*. Contribution of Working Group I to the Third Assessment Report of the Intergovernmental Panel on Climate Change [Houghton, J.T., Y. Ding, D.J. Griggs, M. Noguer, P.J. van der Linden, X. Dai, K. Maskell, and C.A. Johnson (eds.)]. Cambridge University Press, Cambridge, UK, and New York, pp. 471-523.

Min, S.-K. and S. Hense, 2006: A Bayesian approach to climate model evaluation and multi-model averaging with an application to global mean surface temperatures from IPCC AR4 coupled climate models. *Geophysical Research Letters*, **33**, L08708, doi:10.1029/2006GL025779.

Parmesan, C. and G. Yohe, 2003: A globally coherent fingerprint of climate change impacts across natural systems. *Nature*, **421(6918)**, 37-42.

Raftery, A.E., T. Gneiting, F. Balabdaoui, and M. Polakowski, 2005: Using Bayesian model averaging to calibrate forecast ensembles. *Monthly Weather Review*, **133(5)**, 1155-1174.

Salim, A., Y. Pawitan, and K. Bond, 2005: Modelling association between two irregularly observed spatiotemporal processes by using maximum covariance analysis. *Journal of the Royal Statistical Society, Series C*, **54(3)**, 555-573.

Solow, A.R., 2003: Statistics in atmospheric science. *Statistical Science*, **18(4)**, 422-429.

Solow, A.R. and L. Moore, 2000: Testing for a trend in a partially incomplete hurricane record. *Journal Climate*, **13(20)**, 3696-3699.

Solow, A.R. and L. Moore, 2002: Testing for a trend in North Atlantic hurricane activity, 1900-98. *Journal Climate*, **15(21)**, 3111-3114.

Stephenson, D.B., C.A.S. Coelho, F.J. Doblas-Reyes, and M. Balmaseda, 2005: Forecast assimilation: A unified framework for the combination of multi-model weather and climate predictions. *Tellus A*, **57(3)**, 253-264.

Tebaldi, C., L.O. Mearns, D. Nychka, and R.L. Smith, 2004: Regional probabilities of precipitation change: A Bayesian analysis of multimodel simulations. *Geophysical Research Letters*, **31**, L24213, doi:10.1029/2004GL021276.

Tebaldi, C., R.L. Smith, D. Nychka, and L.O. Mearns, 2005: Quantifying uncertainty in projections of regional climate change: A Bayesian approach to the analysis of multimodel ensembles. *Journal Climate*, **18(10)**, 1524-1540.

Vincent, L.A., 1998: A technique for the identification of inhomo-geneities in Canadian temperature series. *Journal of Climate*, **11(5)**, 1094-1104.

Wilke, C.K., L.M. Berliner, and N. Cressie, 1998: Hierarchical Bayesian space-time models. *Environmental and Ecological Statistics*, **5(2)**, 117-154.

Wintle, B.A., M.A. McCarthy, C.T. Volinsky, and R.P. Kavanagh, 2003: The use of Bayesian model averaging to better represent uncertainty in ecological models. *Conservation Biology*, **17(6)**, 1579-1590.

Zwiers, F.W. and H. von Storch, 2004: On the role of statistics in climate research. *International Journal of Climatology*, **24(6)**, 665-680.

PART 5 REFERENCES

Andronova, N. and M.E. Schlesinger, 2001: Objective estimation of the probability distribution for climate sensitivity. *Journal of Geophysical Research*, **106(D19)**, 22605-22612.

Budnitz, R.J., G. Apostolakis, D.M. Boore, L.S. Cluff, K.J. Coppersmith, C.A. Cornell, and P.A. Morris, 1995: *Recommendations for Probabilistic Seismic Hazard Analysis: Guidance on Uncertainty and the Use of Experts.* UCRL-ID 122160. Lawrence Livermore National Laboratory, Livermore CA, 170 pp.

Budnitz, R.J., G. Apostolakis, D.M. Boore, L.S. Cluff, K.J. Coppersmith, C.A. Cornell, and P.A. Morris, 1998: Use of technical expert panels: Applications to probabilistic seismic hazard analysis. *Risk Analysis*, **18(4)**, 463-469.

Cooke, R.M., 1991: *Experts in Uncertainty; Opinion and Subjective Probability in Science.* Oxford University Press, New York, 321 pp.

Dalkey, N.C., 1969: *The Delphi Method: An Experimental Study of the Group Opinion.* RM-5888-PR. Rand Corporation, Santa Monica, CA, 79 pp.

Dalkey, N.C., B. Brown, and S. Cochran, 1970: The use of self-ratings to improve group estimates: Experimental evaluation of Delphi procedures. *Technological Forecasting*, **1(3)**, 283-291.

DeGroot, M., 1970: *Optimal Statistical Decision.* McGraw-Hill, New York, 489 pp.

Evans, J.S., G.M. Gray, R.L. Sielken Jr., A.E. Smith, C. Valdez-Flores, and J.D. Graham, 1994a: Using of probabilistic expert judgment in uncertainty analysis of carcinogenic potency. *Regulatory Toxicology and Pharmacology*, **20(1 pt.1)**, 15-36.

Evans, J.S., J.D. Graham, G.M. Gray, and R.L. Sielken Jr., 1994b: A distributional approach to characterizing low-dose cancer risk. *Risk Analysis*, **14(1)**, 25-34.

Garthwaite, P.H., J.B. Kadane, and A. O' Hagan, 2005: Statistical methods for eliciting probability distributions. *Journal of the American Statistical Association*, **100(470)**, 680-700.

Gritsevskyi, A. and N. Nakićenović, 2000: Modeling uncertainty of induced technological change. *Energy Policy*, **28(13)**, 907-921.

Gustafsen, D.H., R.K. Shukla, A. Delbecq, and G.W. Walster, 1973: A comparative study of group format on aggregative subjective likelihood estimates made by individuals, interactive groups, Delphi groups, and nominal groups. *Organizational Behavior and Human Performance*, **9(2)**, 280-291.

Hammitt, J.K. and A.I. Shlyakhter, 1999: The expected value of information and the probability of surprise. *Risk Analysis*, **19(1)**, 135-152.

IPCC (Intergovernmental Panel on Climate Change), 2005: *Guidance Notes for Lead Authors of the IPCC Fourth Assessment Report on Addressing Uncertainties.* [Intergovernmental Panel on Climate Change, Geneva, Switzerland], 4 pp. Available at: http://www.ipcc.ch/pdf/supporting-material/uncertainty-guidance-note.pdf

IPCC (Intergovernmental Panel on Climate Change), 2007: *Climate Change 2007: The Physical Science Basis.* Contribution of Working Group I to the Fourth Assessment Report of the Intergovernmental Panel on Climate Change [Solomon, S., D. Qin, M. Manning, Z. Chen, M. Marquis, K. Averyt, M.M.B. Tignor, and H. L. Miller (eds.)]. Cambridge University Press, Cambridge, UK, and New York, 996 pp.

Keith, D.W., 1996: When is it appropriate to combine expert judgments? *Climatic Change*, **33(2)**, 139-143.

Kurowicka, D. and R.M. Cooke, 2006: *Uncertainty Analysis with High Dimensional Dependence Modeling.* Wiley, Chichester (UK), 284 pp.

Linstone, H. and M. Turoff (eds.), 1975: *The Delphi Method: Techniques and Applications.* Addison-Wesley, Reading, MA, 618 pp.

McCauley, C., 1989: The nature of social influence in groupthink: Compliance and internalization. *Journal of Personality and Social Psychology*, **57(2)**, 250-260.

Morgan, M.G. and M. Henrion, 1990: *Uncertainty: A Guide to Dealing with Uncertainty in Quantitative Risk and Policy Analysis.* Cambridge University Press, Cambridge, UK, and New York, 332 pp.

Morgan, M.G. and D. Keith, 1995: Subjective judgments by climate experts. *Environmental Science & Technology*, **29(10)**, 468-476.

Morgan, M.G., S.C. Morris, A.K. Meier, and D.L. Shenk, 1978a: A probabilistic methodology for estimating air pollution health effects from coal-fired power plants. *Energy Systems and Policy*, **2**, 287-310.

Morgan, M.G., S.C. Morris, W.R. Rish, and A.K. Meier, 1978b: Sulfur control in coal-fired power plants: A probabilistic approach to policy analysis. *Journal of the Air Pollution Control Association*, **28**, 993-997.

Morgan, M.G., S.C. Morris, M. Henrion, D. Amaral, and W.R. Rish, 1984: Technical uncertainties in quantitative policy analysis: A sulphur air pollution example. *Risk Analysis*, **4(3)**, 201-216.

Morgan, M.G., S.C. Morris, M. Henrion, and D.A.L. Amaral, 1985: Uncertainty in environmental risk assessment: A case study involving sulfur transport and health effects. *Environmental Science & Technology*, **19(8)**, 662-667.

Morgan, M.G., L.F. Pitelka, and E. Shevliakova, 2001: Elicitation of expert judgments of climate change impacts on forest ecosystems. *Climatic Change*, **49(3)**, 279-307.

Morgan, M.G., P.J. Adams, and D. Keith, 2006: Elicitation of expert judgments of aerosol forcing. *Climatic Change*, **75(1-2)**, 195-214.

Moss, R. and S.H. Schneider, 2000: Uncertainties in the IPCC TAR: Recommendations to lead authors for more consistent assessment and reporting. In: *Guidance Papers on the Cross Cutting Issues of the Third Assessment Report of the IPCC* [Pachauri, R., T. Taniguchi, and K. Tanaka (eds.)]. Intergovernmental Panel on Climate Change, Geneva, Switzerland, pp. 33-51. Available at: http://www.ipcc.ch/pdf/supporting-material/guidance-papers-3rd-assessment.pdf

Nakićenović, N. and K. Riahi, 2002: *An Assessment of Technological Change Across Selected Energy Scenarios*. World Energy Council, London, 139 pp.

National Defense University, 1978: *Climate Change to the Year 2000: A Survey of Expert Opinion*. National Defense University, Washington DC, 109 pp.; for a critique see Stewart, T.R. and M.H. Glantz, 1985: Expert judgment and climate forecasting: A methodological critique of climate change to the year 2000. *Climatic Change*, **7(2)**, 159-183.

Nordhaus, W.D., 1994: Expert opinion on climate change. *American Scientist*, **82(1)**, 45-51.

Oppenheimer, M., B.C. O'Neill, M. Webster, and S. Agrawall, 2007: The limits of consensus. *Science*, **317(5844)**, 1505-1506.

Spetzler, C.S. and C.-A.S. Staël von Holstein, 1975: Probability encoding in decision analysis. *Management Science*, **22(3)**, 340-352.

Stewart, T.R., J.L. Mumpower, and P. Reagan-Cirincione, 1992: *Scientists' Agreement and Disagreement About Global Climate Change: Evidence from Surveys*. Center for Policy Research, Nelson A. Rockefeller College of Public Affairs and Policy, State University of New York, Albany, New York, 25 pp. In addition to reporting the results of their e-mail survey, this report also summarizes several other surveys.

von Winterfeldt, D. and W. Edwards, 1986: *Decision Analysis and Behavioral Research*. Cambridge University Press, Cambridge, UK, and New York, 624 pp.

Wallsten, T.S. and R.G. Whitfield, 1986: *Assessing the Risks to Young Children of Three Effects Associated with Elevated Blood Lead Levels*. ANL/AA-32. Argonne National Laboratory, Argonne, IL, 158 pp.

Watson, S.R. and D.M. Buede, 1987: *Decision Synthesis: The Principles and Practice of Decision Analysis*. Cambridge University Press, Cambridge, UK, and New York, 320 pp.

Webster, M., C. Forest, J. Reilly, M. Babiker, D. Kicklighter, M. Mayer, R. Prinn, M. Sarofim, A. Sokolov, P. Stone, and C. Wang, 2003: Uncertainty analysis of climate change and policy response. *Climatic Change*, **61(3)**, 295-350.

Zickfeld, K., A. Levermann, T. Kuhlbrodt. S. Rahmstorf, M.G. Morgan, and D. Keith: 2007: Expert judgements on the response on the Atlantic meridional overturning circulation to climate change. *Climatic Change*, **82(3-4)**, 235-265.

PART 6 REFERENCES

Allen, M., 1999: Do-it-yourself climate prediction. *Nature*, **401(6754)**, 642.

Annan, J.D. and J.C. Hargreaves, 2006: Using multiple observationally-based constraints to estimate climate sensitivity. *Geophysical Research Letters*, **33**, L06704, doi:10.1029/2005GL025259.

Borgonovo, E., 2006: Measuring uncertainty importance: Investigation and comparison of alternative approaches. *Risk Analysis*, **26(5)**, 1349-1361.

Casman, E.A., M.G. Morgan, and H. Dowlatabadi, 1999: Mixed levels of uncertainty in complex policy models. *Risk Analysis*, **19(1)**, 33-42.

Craig, P., A. Gadgil, and J.G. Koomey, 2002: What can history teach us? A retrospective examination of long-term energy forecasts for the United States. *Annual Review of Energy and the Environment*, **27**, 83-118.

Dowlatabadi, H., 2000: Bumping against a gas ceiling. *Climatic Change*, **46(3)**, 391-407.

Dowlatabadi, H. and M.G. Morgan, 1993: A model framework for integrated studies of the climate problem. *Energy Policy*, **21(3)**, 209-221.

Forest, C.E, P.H. Stone, A.P. Sokolov, M.R. Allen, and M.D. Webster, 2002: Quantifying uncertainties in climate system properties with the use of recent climate observations. *Science*, **295(5552)**, 113-116.

Forest, C.E, P.H. Stone, and A.P. Sokolov, 2006: Estimated PDF's of climate system properties including natural and anthropogenic forcings. *Geophysical Research Letters*, **33**, L01705, doi:10.1029/2005GL023977.

Frame, D.J., B.B.B. Booth, J.A Kettleborough, D.A. Stainforth, J.M. Gregory, M. Collins, and M.R. Allen, 2005: Constraining climate forecasts: The role of prior assumptions. *Geophysical Research Letters*, **32**, L09702, doi:10.1029/2004GL022241.

IPCC (Intergovernmental Panel on Climate Change), 2001: *Climate Change 2001: The Scientific Basis*. Contribution of Working Group I to the Third Assessment Report of the Intergovernmental Panel on Climate Change [Houghton, J.T., Y. Ding, D.J. Griggs, M. Noguer, P.J. van der Linden, X. Dai, K. Maskell, and C.A. Johnson (eds.)]. Cambridge University Press, Cambridge, UK, and New York, 881 pp.

Keller, K., M. Hall, S.-R. Kim, D.F. Bradford, and M. Oppenheimer, 2005: Avoiding dangerous anthropogenic interference with the climate system. *Climatic Change*, **73(3)**, 227-238.

Morgan, M.G. and H. Dowlatabadi, 1996: Learning from integrated assessment of climate change. *Climatic Change*, **34(3-4)**, 337-368.

Morgan, M.G. and M. Henrion, 1990: *Uncertainty: A Guide to Dealing with Uncertainty in Quantitative Risk and Policy Analysis*. Cambridge University Press, Cambridge, UK, and New York, 332 pp.

Murphy, J.M., D.M.H. Sexton, D.N. Barnett, G.S. Jones, M.J. Webb, M. Collins, and D.A. Stainforth, 2004: Quantification of modeling uncertainties in a large ensemble of climate change simulations. *Nature*, **430(7001)**, 768-772.

Nordhaus, W.D. and J. Boyer, 2000: *Warming the World: Economic Models of Global Warming*. MIT Press, Cambridge, MA, 232 pp.

Roe, G.H. and M.B. Baker, 2007: Why is climate sensitivity so unpredictable? *Science*, **318(5850)**, 629-632.

Smil, V., 2003: *Energy at the Crossroads*. MIT Press, Cambridge, MA, 448 pp.

Sokolov, A.P., C.A. Schlosser, S. Dutkiewicz, S. Paltsev, D.W. Kicklighter, H.D. Jacoby, R.G. Prinn, C.E. Forest, J. Reilly, C. Wang, B. Felzer, M.C. Sarofim, J. Scott, P.H. Stone, J.M.

Melillo, and J. Cohen, 2005: *The MIT Integrated Global System Model (IGSM) Version 2: Model Description and Baseline Evaluation*. MIT Global Change Joint Program report 124. Joint Program on the Science and Policy of Global Change, Cambridge, MA, 40 pp. Available at http://globalchange.mit.edu/pubs/all-reports.php

Stainforth, D.A., T. Aina, C. Christensen, M. Collins, N. Faull, D.J. Frame, J.A. Ketteborough, S. Knight, A. Martin, J.M. Murphy, C. Piani, D. Sexton, L.A. Smith, R.A. Spicer, A.J. Thorpe, and M.R. Allen, 2005: Uncertainty in predictions of the climate response to rising levels of greenhouse gases. *Nature*, **433(7024)**, 403-406.

Webster, M., C. Forest, J. Reilly, M. Babiker, D. Kicklighter, M. Mayer, R. Prinn, M. Sarofim, A. Sokolov, P. Stone, and C. Wang, 2003: Uncertainty analysis of climate change and policy response. *Climatic Change*, **61(3)**, 295-350.

PART 7 REFERENCES

Ben-Haim, Y., 2001: *Information-Gap Decision Theory: Decisions Under Severe Uncertainty*. Academic Press, San Diego, CA, 330 pp.

Casman, E.A., M.G. Morgan, and H. Dowlatabadi, 1999: Mixed levels of uncertainty in complex policy models. *Risk Analysis*, **19(1)**, 33-42.

Clark, W.C., 1980: Witches, floods and wonder drugs: Historical perspectives on risk management. In *Societal Risk Assessment: How Safe is Safe Enough?* [Schwing, R.C. and W.A. Albers Jr. (eds.)]. Plenum, New York, pp. 287-313.

DeKay, M.L., M.J. Small, P.S. Fischbeck, R.S. Farrow, A. Cullen, J.B. Kadane, L.B. Lave, M.G. Morgan, and K. Takemura, 2002: Risk-based decision analysis in support of precautionary policies. *Journal of Risk Research*, **5(4)**, 391-417.

Dessai, S. and M. Hulme, 2007: Assessing the robustness of adaptation decisions to climate change uncertainties: A case-study on water resources management in the East of England. *Global Environmental Change*, **17(1)**, 59-72.

Dowlatabadi, H., 1998: Sensitivity of climate change mitigation estimates to assumptions about technical change. *Energy Economics*, **20(5-6)**, 473-493.

Fischhoff, B., 1991: Value elicitation: Is there anything in there? *American Psychologist*, **46(8)**, 835-847.

Fischhoff, B., 2005: Cognitive processes in stated preference methods. In: *Handbook of Environmental Economics* [Mäleer, K.-G. and J.R. Vincent (eds.)]. Elsevier, Amsterdam and Boston, MA, volume 2, pp. 938-968.

Glantz, M.H., D.G. Streets, T.R. Stewart, N. Bhatti, C.M. Moore, and C.H. Rosa, 1998: *Exploring the Concept of Climate Surprises: A Review of the Literature on the Concept of Surprise and How it Relates to Climate Change.* Environment and Social Impacts Group and the National Center for Atmospheric Research, Boulder, CO, 88 pp.

Groves, D.G., 2006: *New Methods for Identifying Robust Long-term Water Resources Management Strategies for California.* Ph.D. thesis. RAND Graduate School, Santa Monica, CA, 206 pp.

Hammond, J.S., R.L. Keeney, and H. Raiffa, 1999: *Smart Choices: A Practical Guide to Making Better Decisions.* Harvard Business School Press, Boston, MA, 244 pp.

Hastie, R. and R.M. Dawes, 2001: *Rational Choice in an Uncertain World: The Psychology of Judgment and Decision Making.* Sage, Thousand Oaks, CA, 372 pp.

Hine, D. and J.W. Hall, 2007: Analysing the robustness of engineering decisions to hydraulic model and hydrological uncertainties. In: *Harmonizing the Demands of Art and Nature in Hydraulics.* Proceedings of 32nd Congress of IAHR (International Association of Hydraulic Engineering and Research), Venice, July 1-6. CORILA, Venice, Italy, 791 pp.

Hollings, C.C., 1986: The resilience of terrestrial ecosystems: Local surprise and global change. In: *Sustainable Development in the Biosphere,* [Clark, W.C. and R.E. Munn (eds.)]. Cambridge University Press, Cambridge, UK, and New York, 491 pp.

Howard, R.A. and J.E. Matheson (eds.), 1977: *Readings in Decision Analysis.* Decision Analysis Group, SRI International, Menlo Park, CA, 2nd edition, 613 pp.

IPCC (Intergovernmental Panel on Climate Change), 2001: *Climate Change 2001: Mitigation.* Contribution of Working Group III to the Third Assessment Report of the Intergovernmental Panel on Climate Change [Metz, B., O. Davidson, R. Swart, and J. Pan (eds.)]. Cambridge University Press, Cambridge, UK, and New York, 881 pp.

IPCC (Intergovernmental Panel on Climate Change), 2007: *Climate Change 2007: The Physical Science Basis.* Contribution of Working Group I to the Fourth Assessment Report of the Intergovernmental Panel on Climate Change [Solomon, S., D. Qin, M. Manning, Z. Chen, M. Marquis, K. Averyt, M.M.B. Tignor, and H.L. Miller (eds.)]. Cambridge University Press, Cambridge, UK, and New York, 996 pp.

Jaeger, C., O. Renn, E.A. Rosa, and T. Webler, 1998: Decision analysis and rational action. In: *Human Choice and Climate Change, Vol. 3: Tools for Policy Analysis* [Rayner, S. and E.L. Malone (eds.)]. Battelle Press, Columbus, OH, pp. 141-215.

Kahneman, D., P. Slovic, and A. Tversky (eds.), 1982: *Judgment Under Uncertainty: Heuristics and Biases.* Cambridge University Press, Cambridge, UK, and New York, 551 pp.

Keeney, R.L., 1982: Decision analysis: An overview. *Operations Research,* **30(5)**, 803-837.

Keeney, R.L., 1992: *Value-Focused Thinking: A Path to Creative Decision Making.* Harvard University Press, Cambridge, MA, 416 pp.

Knight, F.H., 1921: *Risk, Uncertainty and Profit.* Houghton Mifflin Company, Boston, MA, 381 pp.

La Porte, T.R., 1996: High reliability organizations: Unlikely, demanding, and at risk. *Journal of Contingencies and Crisis Management,* **63(4)**, 60-71.

La Porte, T.R. and P.M. Consolini, 1991: Working in practice but not in theory: Theoretical challenges of high-reliability organizations. *Journal of Public Administration Research and Theory: J-PART,* **1(1)**, 19-48.

Lee, K., 1993: *Compass and Gyroscope: Integrating Science and Politics for the Environment.* Island Press, Washington, DC, 243 pp.

Lempert, R.J., M.E. Schlesinger, and S.C. Bankes, 1996: When we don't know the costs or the benefits: Adaptive strategies for abating climate change. *Climatic Change,* **33(2)**, 235-274.

Lempert, R.J., M.E. Schlesinger, S.C. Bankes, and N.G. Andronova, 2000: The impact of variability on near-term climate-change policy choices and the value of information. *Climatic Change,* **45(1)**, 129-161.

Lempert, R.J., S.W. Popper, and S.C. Bankes, 2002: Confronting surprise. *Social Science Computing Review,* **20(4)**, 420-440.

Lempert, R.J., S.W. Popper, and S.C. Bankes, 2003: *Shaping the Next One Hundred Years: New Methods for Quantitative, Long-term Policy Analysis.* MR-1626-RPC. RAND Corporation, Santa Monica, CA, 187 pp.

Lempert, R.J. and M.E. Schlesinger, 2006: Adaptive strategies for climate change. In: *Innovative Energy Strategies for CO_2 Stabilization* [R.G. Watts (ed.)]. Cambridge University Press, Cambridge, UK, and New York, pp. 45-86.

Lerner, J.S. and L.Z. Tiedens, 2006: Portrait of the angry decision maker: How appraisal tendencies shape anger influence on decision making. *Journal of Behavioral Decision Making,* **19(2)**, 115-137.

Lerner, J.S., R.M. Gonzalez, D.A. Small, and B. Fischhoff, 2003: Effects of fear and anger on perceived risks of terrorism. *Psychological Science,* **14(2)**, 144-150.

Levin, A.T. and J.C. Williams, 2003: Robust monetary policy with competing reference models. *Journal of Monetary Economics,* **50(5)**, 945-975.

Loewenstein, G.F., E.U. Weber, C.K. Hsee, and E.S. Welch, 2001: Risk as feelings. *Psychological Bulletin*, **127(2)**, 267-286.

Morgan, M.G. and H. Dowlatabadi, 1996: Learning from integrated assessment of climate change. *Climatic Change*, **34(3-4)**, 337-368.

Morgan, M.G. and D. Keith, 1995: Subjective judgments by climate experts. *Environmental Science & Technology*, **29(10)**, 468-476.

Morgan, M.G., M. Kandlikar, J. Risbey, and H. Dowlatabadi, 1999: Why conventional tools for policy analysis are often inadequate for problems of global change. *Climatic Change*, **41(3-4)**, 271-281.

Morgan, M.G., P.J. Adams, and D. Keith, 2006: Elicitation of expert judgments of aerosol forcing. *Climatic Change*, **75(1-2)**, 195-214.

NAST (National Assessment Synthesis Team), 2001: *Climate Change Impacts on the United States: The Potential Consequences of Climate Variability and Change.* Cambridge University Press, Cambridge, UK, and New York, 612 pp.

NRC (National Research Council), 1986: *Understanding Risk: Informing Decisions in a Democratic Society.* National Academy Press, Washington, DC, 249 pp.

NRC (National Research Council), 2002: *Abrupt Climate Change: Inevitable Surprises.* National Academy Press, Washington, DC, 244 pp.

Parson, E., V. Burkett, K. Fisher-Vanden, D. Keith, L. Mearns, H. Pitcher, C. Rosenzweig, and M. Webster, 2007. *Global Change Scenarios: Their Development and Use.* Sub-report 2.1B of Synthesis and Assessment Product 2.1 by the U.S. Climate Change Science Program and the Subcommittee on Global Change Research. Department of Energy, Office of Biological & Environmental Research, Washington, DC, 106 pp.

Paté-Cornell, M.E., L.M. Lakats, D.M. Murphy, and D.M. Gaba, 1997: Anesthesia patient risk: A quantitative approach to organizational factors and risk management options. *Risk Analysis*, **17(4)**, 511-523.

Peters, E., D. Västfjäl, T. Gärling, and P. Slovic, 2006: Affect and decision making: A hot topic. *Journal of Behavioral Decision Making*, **19(2)**, 79-85.

Pool, R., 1997: Managing the Faustian bargain. In: *Beyond Engineering: How Society Shapes Technology.* Oxford University Press, New York, pp. 249-277.

Raiffa, H. and R. Schlaifer, 1968: *Applied Statistical Decision Theory.* MIT Press, Cambridge, MA, 356 pp.

Rosenhead, J., 2001: Robustness analysis: Keeping your options open. In: *Rational Analysis for a Problematic World Revisited: Problem Structuring Methods for Complexity, Uncertainty, and Conflict* [Rosenhead, J. and J. Mingers, (eds.)]. Wiley and Sons, Chichester, UK, 2nd edition, pp. 181-208.

Rosenhead, J. and J. Mingers, 2001: *Rational Analysis for a Problematic World Revisited: Problem Structuring Methods for Complexity, Uncertainty, and Conflict.* Wiley and Sons, Chichester, UK, 2nd edition, 366 pp.

Schelling, T.C., 1995: Intergenerational discounting. *Energy Policy*, **23(4-5)**, 395-402.

Schneider, S.H., B.L. Turner, and H. Morehouse Garriga, 1998: Imaginable surprise in global change science. *Journal of Risk Research*, **1(2)**, 165-185.

Schneller, G.O. and G.P. Sphicas, 1983: Decision making under uncertainty: Starr's Domain criterion. *Theory and Decision*, **15(4)**, 321-336.

Simon, H., 1959: Theories of decision-making in economic and behavioral science. *American Economic Review*, **49(3)**, 553-283.

Slovic, P., M.L. Finucane, E. Peters, and D.G. MacGregor, 2004: Risk as analysis and risk as feelings: Some thoughts about affect, reason, risk and rationality. *Risk Analysis*, **24(2)**, 311-322.

Sunstein, C.R., 2005: *Laws of Fear: Beyond the Precautionary Principle.* Cambridge University Press, Cambridge, UK, and New York, 234 pp.

Thompson, M., 1980: Aesthetics of risk: Culture or context. In: *Societal Risk Assessment: How Safe is Safe Enough?* [Schwing, R.C. and W.A. Albers (eds.)]. Plenum, New York, pp. 273-285.

van Notten, W.F., A.M. Sleegers, and A. van Asselt, 2005: The future shocks: On discontinuity and scenario development. *Technological Forecasting and Social Change*, **72(2)**, 175-194.

Vaughan, D., 1996: *The Challenger Launch Decision: Risky Technology, Culture and Deviance at NASA.* University of Chicago Press, Chicago, IL, 575 pp.

Vercelli, A., 1994: *Hard Uncertainty and the Environment.* Nota di Lavoro 46.94. Fondazione ENI Enrico Mattei (FEEM), Milan, Italy.

von Winterfeldt, D. and W. Edwards, 1986: *Decision Analysis and Behavioral Research.* Cambridge University Press, Cambridge, UK, and New York, 624 pp.

Watson, S.R. and D.M. Buede, 1987: *Decision Synthesis: The Principles and Practice of Decision Analysis*. Cambridge University Press, Cambridge, UK, and New York, 320 pp.

Weick, K.E. and K.M. Sutcliffe, 2001: *Managing the Unexpected: Assuring High Performance in an Age of Complexity*. Jossey-Bass, San Francisco, CA, 200 pp.

Wiener, J.B. and M.D. Rogers, 2002: Comparing precaution in the United States and Europe. *Journal of Risk Research*, **5(4)**, 317-349.

Wildavsky, A., 1979: No risk is the highest risk of all. *American Scientist*, **67**, 32-37.

Wilbanks, T. and R. Lee, 1985: Policy analysis in theory and practice. In: *Large-Scale Energy Projects: Assessment of Regional Consequences* [Lakshmanan, T.R. and B. Johansson (eds.)]. North-Holland, Amsterdam and New York, pp. 273-303.

PART 8 REFERENCES

Bostrom, A., M.G. Morgan, B. Fischhoff, and D. Read, 1994: What do people know about global climate change? Part 1: Mental models. *Risk Analysis*, **14(6)**, 959-970.

Dawes, R.M., 1988: *Rational Choice in an Uncertain World*. Harcourt Brace Jovanovich, San Diego, CA, 346 pp.

Fischhoff, B., A. Bostrom, and M. Jacobs-Quadrel, 2002: Risk perception and communication. In: *Oxford Textbook of Public Health* [Detels, R., J. McEwen, R. Reaglenhole, and H. Tanaka (eds.)]. Oxford University Press, New York, 4th edition, pp. 1105-1123.

Friedman, S.M., S. Dunwoody, and C.L. Rogers, 1999: *Communicating Uncertainty: Media Coverage of New and Controversial Science*. L. Erlbaum, Mahwah, NJ, 277 pp.

Ibrekk, H. and M.G. Morgan, 1987: Graphical communication of uncertain quantities to nontechnical people. *Risk Analysis*, **7(4)**, 519-529.

Kempton, W., 1991: Lay perspectives on global climate change. *Global Environmental Change*, **1(3)**, 183-208.

Kempton, W., J.S. Boster, and J.A. Hartley, 1995: *Environmental Values in American Culture*. MIT Press, Cambridge, MA, 320 pp.

Morgan, M.G. and M. Henrion, 1990: *Uncertainty: A Guide to Dealing with Uncertainty in Quantitative Risk and Policy Analysis*. Cambridge University Press, Cambridge, UK, and New York, 332 pp.

Morgan, M.G. and T. Smuts 1994: *Global Warming and Climate Change*, 9 pp. A hierarchically organized brochure with three supporting brochures: *Details Booklet Part 1: More on What is Climate Change?*, 9 pp.; *Details Booklet Part 2: More on If Climate Changes What Might Happen?*, 9 pp.; and *Details Booklet Part 3: More on What can be Done about Climate Change?*, 14 pp. Department of Engineering and Public Policy, Carnegie Mellon University.

Morgan, M.G., B. Fischhoff, A. Bostrom, and C. Atman, 2002: *Risk Communication: A Mental Models Approach*. Cambridge University Press, New York, 351 pp.

OFCM (Office of the Federal Coordinator for Meteorology) 2001: *Toward a Safer America: Building Natural Hazard Resistant Communities through Risk Management and Assessments*. Proceedings of the Forum on Risk Management and Assessments of Natural Hazards, February 5-6, 2001, Washington, D.C. Office of the Federal Coordinator for Meteorological Services and Supporting Research, Washington, DC, 252 pp. Available at: http://www.ofcm.gov/risk/proceedings/riskproceedings2001.htm

Patt, A.G. and D.P. Schrag, 2003: Using specific language to describe risk and probability. *Climatic Change*, **61(1-2)**, 17-30.

Read, D., A. Bostrom, M.G. Morgan, B. Fischhoff, and T. Smuts, 1994: What do people know about global climate change? Part 2: Survey studies of educated laypeople. *Risk Analysis*, **14(6)**, 971-982.

Reiner, D.M., T.E. Curry, M.A. deFigueiredo, H.J. Herzog, S.D. Ansolabehere, K. Itaoka, F. Johnsson, and M. Odenberger, 2006: American exceptionalism? Similarities and differences in the national attitudes toward energy policy and global warming. *Environmental Science & Technology*, **40(7)**, 2093-2098.

PART 9 REFERENCES

Casman, E.A., M.G. Morgan, and H. Dowlatabadi, 1999: Mixed levels of uncertainty in complex policy models. *Risk Analysis*, **19(1)**, 33-42.

Funtowicz, S.O. and J.R. Ravetz, 1990: *Uncertainty and Quality in Science for Policy*. Kluwer Academic Publishers, Dordrecht, the Netherlands, 229 pp.

IPCC (Intergovernmental Panel on Climate Change), 2001: *Climate Change 2001: The Scientific Basis*. Contribution of Working Group I to the Third Assessment Report of the Intergovernmental Panel on Climate Change [Houghton, J.T., Y. Ding, D.J. Griggs, M. Noguer, P.J. van der Linden, X. Dai, K. Maskell, and C.A. Johnson (eds.)]. Cambridge University Press, Cambridge, UK, and New York, 881 pp.

Morgan, M.G. and M. Henrion, 1990: *Uncertainty: A Guide to Dealing with Uncertainty in Quantitative Risk and Policy Analysis*. Cambridge University Press, Cambridge, UK, and New York, 332 pp.

NAST (National Assessment Synthesis Team), 2001: *Climate Change Impacts on the United States: The Potential Consequences of Climate Variability and Change*. Cambridge University Press, Cambridge, UK, and New York, 612 pp.

Smil, V., 2003: *Energy at the Crossroads*. MIT Press, Cambridge, MA, 448 pp.

Tukey, J.W., 1977: *Exploratory Data Analysis*. Addison-Wesley, Boston, MA, 688 pp.

Contact Information

Global Change Research Information Office
c/o Climate Change Science Program Office
1717 Pennsylvania Avenue, NW
Suite 250
Washington, DC 20006
202-223-6262 (voice)
202-223-3065 (fax)

The Climate Change Science Program incorporates the U.S. Global Change Research Program and the Climate Change Research Initiative.

To obtain a copy of this document, place an order at the Global Change Research Information Office (GCRIO) web site: http://www.gcrio.org/orders.

Climate Change Science Program and the Subcommittee on Global Change Research

William Brennan, Chair
Department of Commerce
National Oceanic and Atmospheric Administration
Director, Climate Change Science Program

Jack Kaye, Vice Chair
National Aeronautics and Space Administration

Allen Dearry
Department of Health and Human Services

Anna Palmisano
Department of Energy

Mary Glackin
National Oceanic and Atmospheric Administration

Patricia Gruber
Department of Defense

William Hohenstein
Department of Agriculture

Linda Lawson
Department of Transportation

Mark Myers
U.S. Geological Survey

Tim Killeen
National Science Foundation

Patrick Neale
Smithsonian Institution

Jacqueline Schafer
U.S. Agency for International Development

Joel Scheraga
Environmental Protection Agency

Harlan Watson
Department of State

EXECUTIVE OFFICE AND OTHER LIAISONS

Robert Marlay
Climate Change Technology Program

Katharine Gebbie
National Institute of Standards & Technology

Stuart Levenbach
Office of Management and Budget

Margaret McCalla
Office of the Federal Coordinator for Meteorology

Robert Rainey
Council on Environmental Quality

Daniel Walker
Office of Science and Technology Policy

www.ingramcontent.com/pod-product-compliance
Lightning Source LLC
Chambersburg PA
CBHW080643180526

45168CB00008B/3285

* 9 7 8 1 5 0 7 8 0 9 3 7 2 *